Pest management

G. A. Matthews

*Reader in Pest Management,
Department of Pure and
Applied Biology – and Director,
Silwood Centre for Pest Management,
Imperial College, London.*

Longman
London and New York

Longman Group Limited
Longman House, Burnt Mill, Harlow
Essex CM20 2JE, England
Associated companies throughout the world

Published in the United States of America
by Longman Inc., New York

First published 1984

British Library Cataloguing in Publication Data
Matthews, G. A.
Pest management.
1. Pest control
I. Title
632′.9 SB950

ISBN 0-582-47011-0

Library of Congress Cataloging in Publication Data
Matthews, G. A.
Pest management.

Includes bibliographical references and index.
1. Pest control. I. Title.
SB950.M39 1984 632′.9 83.11976
ISBN 0-582-47011-0

Set in 10/12pt Linotron 202 Times Roman
Printed in Singapore by
The Print House (Pte) Ltd

Contents

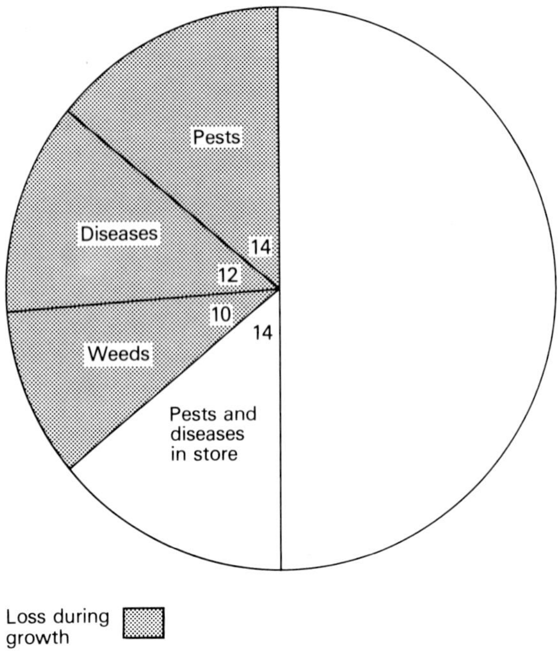

Estimated losses in crops (percentage of potential yield) (from World (FAO) statistics after Cramer 1967)

Preface

It has been estimated that agricultural pests in the form of insects, diseases and weeds account for the loss of about one-half of the world's food supply; in addition, many other pests spread diseases of man. In his struggle against them, man has relied increasingly on the use of chemical pesticides, but concern about his future environment has stimulated the search for more ecologically orientated methods of controlling pest populations, and this has led directly to the concept of *pest management*.

Pest management has been defined by the Food and Agriculture Organization (FAO) as a system than in the context of the associated environment and the population dynamics of the pest species, utilizes all suitable techniques and methods in as compatible a manner as possible and maintains the pest populations at levels below those causing economic injury.

Various authors have described individual control tactics involving host resistance, biological control, cultural control or chemical control, and have given examples where they have been successfully integrated. This book is for those concerned with the development of a pest management system and aims to provide a better understanding of the techniques and equipment needed before such a programme can be implemented. Different techniques that can be used to assess the importance of a pest and evaluate the role of individual tactics are described. As pesticides remain an important tool for the rapid reduction of a pest population, the processes of evaluating pesticides in the laboratory and field are considered in more detail. Finally, the way in which appropriate tactics may be integrated into a coordinated regional pest management programme is discussed.

Acknowledgements

I wish to thank my wife, Moira, for her encouragement in the preparation of this book and all those who have provided valuable assistance: in particular I thank Dr John Mumford (Ch. 1); Dr John Galley and Dr John Stevenson (Ch. 5); Edward Bals (Ch. 6); Dr Michael Crawley (Ch. 7); Dr Geoffrey Norton (Ch. 8); Dr Walter Jepson, OBE and Gordon Wrigley who read through several chapters; and Mr Ray Treichler for helping to obtain some information from the USA. I thank Mrs Patricia Grellier for typing the book.
A number of photographs (Figs 2.3, 2.6(c), 2.7(c) and 2.9) used in the text have been taken at the Overseas Spraying Machinery Centre by Frank Wright.

I am responsible for views expressed in the book and hope that the information on a number of aspects of pest management brought together here will assist students and others develop more efficient ways of regulating pest populations without harmful effects on the environment. I will be most grateful for any comments and corrections provided by readers.

Note: The author has endeavoured to ascertain the accuracy of the statements in this book. However, facilities for determining such accuracy with absolute certainty in relation to every particular statement have not necessarily been available. Therefore the reader is advised to seek an independent expert opinion before implementing in practice any particular technique described herein.

The publishers are grateful to the following for permission to reproduce copyright material:

The author, Dr. Akobundu for our Table 1.7 (Carson 1978); Edward Arnold Ltd for our figs 5.12, 5.13, 5.15, 5.16 (Hewlett & Plackett 1979) & our Table 2.4 from Table 6 *Diseases in Crops* by B.E.J. Wheeler in 'Studies in Biology' Series; Associated Book Publishers Ltd for our figs 2.4, 2.5, 2.6, 2.7(a) & (b), 2.10, 2.12 (Southwood 1966); Blackwell Scientific Publications Ltd for our Table 1.6 (George & Gair 1979); British Crop Protection Council for our Tables 1.8 (Clare, Boothroyd & Simkin 1981), 2.5, 2.6 (Way & Cammell 1977, 1979), 4.4 from Table 7.16 *Pest & Disease Control Handbook* 1st edition, 6.3 (Lawson & UK 1979); Butterworth Scientific for our Tables 8.2, 8.3, 8.4 from p 147 by Dr. G. Norton *Crop Protection* Vol. 1; Cambridge University Press Ltd for our Tables 1.1 (Way 1961), 1.9 (Matthews, Rowell & Beeden 1972); Commonwealth Agricultural Bureaux for our fig 3.5 (Marks 1977), our Tables 2.2, 2.3 (Marks 1976b & 1978), 5.2 (Matthews 1966); Commonwealth Agricultural Bureaux & the author, Prof. J.R. Busvine for our figs 5.1, 5.2, 5.3, 5.8, 5.9, 5.10 (Busvine 1971); East Malling Research Station for our Table 8.5 (Cranham & Solomon 1981); Food & Agriculture Organisation of the United Nations for our Table 1.3 (Church 1971); The Controller of Her Britannic Majesty's Stationery Office & the author, C.J. Stent for our fig 6.7 from fig 1 (Stent, Taylor & Shaw 1981); The International Rice Research Institute for our Tables 2.1 (Hirao 1979), 7.1 (Gomez 1972); Longman Group Ltd for our Tables 6.4, 6.5, 6.6 (Matthews 1979a), Appendix 1 Table $2Y_o$, Appendix 2 'Random Numbers' (Fisher & Yates 1974); Macmillan Journals Ltd for our Table 1.10 (Conway 1977) Copyright © Macmillan Journals Ltd; Natural Environment Research Council for our Table 4.1 (Elliott & Graham-Bryce 1977); Rothamsted Experimental Station for our Table 4.3 (Stevenson, Needham & Walker 1978); The Royal Society & the author, I.J. Graham-Bryce for our Table 6.2 (Graham-Bryce 1981); the author, Dr. Nigel E.A. Scopes for our fig 3.2 (Scopes, Randall & Biggerstaff 1975).

List of abbreviations

CDA	Controlled droplet application
CV	Coefficient of variation
EIL	Economic injury level
ELISA	Enzyme-linked immunosorbent assay
EP	Equilibrium position
EPA	Environmental Protection Agency
ET	Economic threshold
FAO	Food and Agriculture Organization
IOBC	International Organization for Biological Control of Noxious Animals and Plants
IPM	Integrated pest management
LSD	Least significant difference
NMD	Number median diameter
MCPA	4-chloro-2-methylphenoxyacetic acid
MCPB	4-(4-chloro-2-methylphenoxy) butanoic acid
OPS	Oxford Precision Sprayer
RH	Relative humidity
SE	Standard error
ULV	Ultra low volume
VMD	Volume median diameter
WHO	World Health Organization
WRO	Weed Research Organization

1

What is a pest?

A pest has been defined as a living organism which causes damage or illness to man or his possessions, or is otherwise in some sense 'not wanted' (Conway 1976). Organisms which can be pests include insects, mites, ticks, nematodes, fungi, bacteria, weeds, rodents, birds, molluscs, crustaceans and viruses. Whether an organism is a pest or not basically depends on a value judgement. It may not be a pest in its natural habitat, but its status can be changed when it comes into conflict with man; thus, for example, the frit fly *Oscinella frit* may not be a pest in grasslands but it becomes one when crops of oats or maize are attacked. This change in status can be very noticeable when man practises monoculture, i.e. provides large areas of a single crop.

Similarly, weeds are plants out of place which compete with agricultural crops for nutrients, water and light, especially in the early stages of crop establishment, or interfere with harvesting or lower the quality of a crop. Some crop species themselves may become serious weeds when they remain after harvest and compete with the subsequent crops of a rotation.

Farmers will naturally take action if pests reduce their revenue, though even when pests are absent control decisions may be taken to reduce the risk of infestation (Mumford 1981). Some farmers may respond to the risk of uncertain pest attack by a schedule of prophylactic treatments in which the timing and number are quite independent of actual pest numbers. The decision is therefore normally taken irrespective of whether the pest will arrive or not.

Sometimes a pesticide is applied, not because it is obviously necessary, but because the farmer takes the opportunity to be 'safe' for little extra effort. Such opportunistic treatment occurs particularly when a farmer has to apply a spray anyway, e.g. a fungicide on deciduous fruit for mildew control, and just adds an insecticide to save a separate application later.

The increased use of pesticides as a result of prophylactic and opportunistic treatments may not always be economically optimal or ecologically acceptable.

Ideally, treatments should be based on the estimated losses in revenue in relation to estimated pest populations. The effect of one particular pest may be of crucial importance in a crop or ecological area, e.g. the devastation that can be caused by a swarm of locusts. More frequently damage is the cumulative results of several pests. In the apple orchard, the farmer has to consider codling and tortricid larvae attacking the fruit, aphids and mites on leaves, diseases including mildew and scab and also weeds. Instead of studying individual pests, more attention is now being given to the interaction between two or more organisms in a pest complex. In addition to the direct damage caused by one organism, the attacked plant's susceptibility to an additional pest may be increased; thus, for example, *Fusarium* and *Verticillium* wilt fungi tend to have more serious effects after nematode damage.

A pest which causes major damage every season unless controlled is regarded as a *key pest*, whereas those causing damage at irregular intervals are known as *occasional pests*. Many organisms are potential pests, but only become so if an important controlling factor is disrupted.

A pest management system can be developed to meet immediate requirements for a practical control programme without detailed ecological studies, but to improve control strategies a thorough knowledge of pest population levels and population dynamics is needed (Way 1973). General equilibrium position (EP) is the average population density of a pest population over a long period of time, unaffected by temporary interventions of pest control. The pest population will fluctuate about this equilibrium according to the influence of density-dependent factors such as parasitoids, predators and diseases. The *economic injury level* (EIL) was defined by Headley (1972a, b) as the population that produces incremental damage equal to the cost of preventing damage, i.e. the lowest pest population which will cause economic damage (Stern *et al.* 1959). The *economic threshold* (ET) is the population density at which control measures should be determined to prevent an increasing pest population from reaching the EIL. The ET is lower than the EIL so as to give sufficient time to put control measures into operation and have them take effect.

Figure 1.1 illustrates the relative positions of the EIL, ET and EP for a non-pest (A), an occasional pest (B), a perennial pest (C) and a severe pest (D). A perennial or severe pest may also be a key pest.

The importance of any pest must be related to the value of damage it causes. Damage may be quantifiable – e.g. a reduction in yield or the number of animals infected with parasites – though qualitative judgements are often necessary. Thus lower grade of produce or simply a delay in harvesting can have a considerable impact on the price paid, particularly with many vegetable crops that undergo price fluctuations.

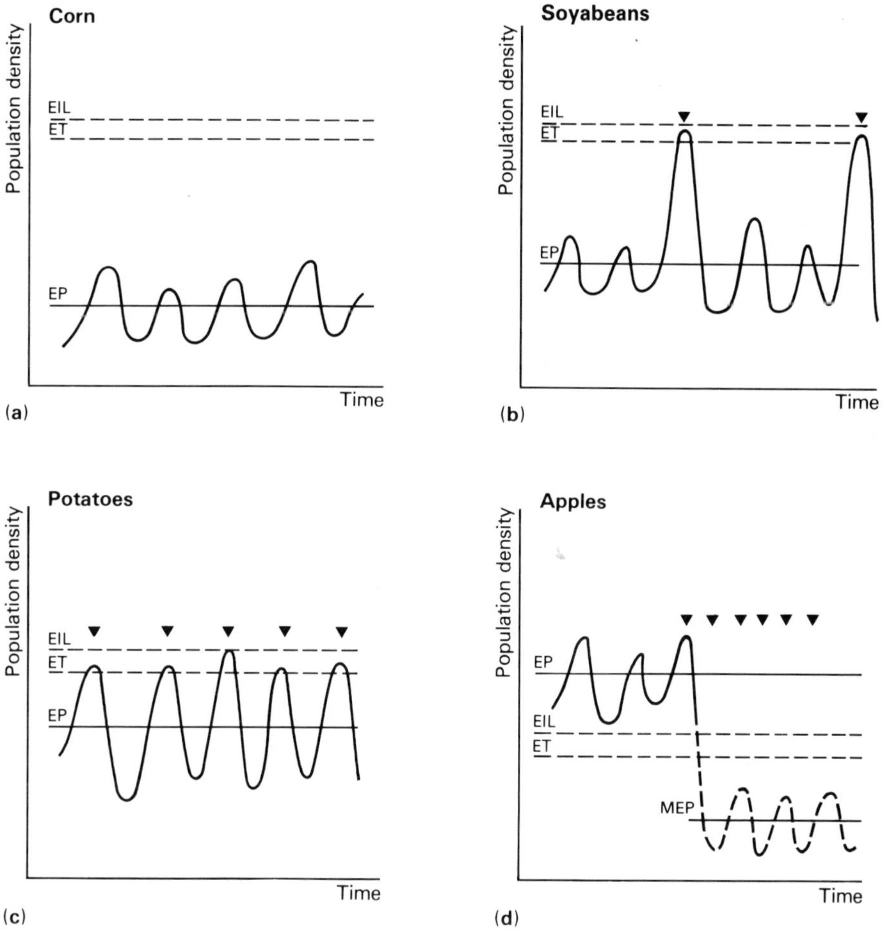

EIL — economic injury level
ET — economic threshold
EP — equilibrium position
MEP — modified equilibrium position
▼ — pest control intervention

Fig. 1.1 Economic injury levels and economic thresholds for typical insect pest situations:
(a) Yellow Woolly Bear, *Diacrisia virginica*
(b) Green Cloverworm, *Plathypena scobra*
(c) Colorado Potato Beetle, *Leptinotarsa decemlineata*
(d) Codling Moth, *Cydia pomonella*
(Modified from Metcalf and Luckmann 1975).

Types of damage

Direct damage

The most direct losses are incurred when pests completely destroy plants or damage those parts of plants which are harvested. Many examples could be quoted, but a few have been selected to indicate the extent to which damage can occur.

Complete loss of crops to the locusts has already been mentioned, but equally catastrophic damage can be caused by armyworm larvae, leaf-cutting ants, termites and other insect pests. Rats also cut down plants by gnawing at their bases. Diseases such as blister blight of tea can cause devastating losses; young seedlings can be killed by damping-off and seedling blights, although such severe damage may be localized.

Fruit and seeds of many crops are partially or totally damaged by various lepidopteran larvae, birds and rats. Diseases such as apple scab or common scab of potatoes significantly reduce the market value and storage quality of the harvested produce. Leaf feeders are direct pests on a crop such as lettuce, but not on brussels sprouts; similarly root pests cause direct damage to cassava, but indirect damage to maize.

Indirect damage

Damage to parts of plants other than product

Often damage is caused to leaves and other parts of plants which reduces the ability of the plant to photosynthesize and transfer the products to the fruit, tubers or other harvested part. Plants may naturally produce an excess of photosynthetic products, so some losses can be tolerated without affecting yield. Some plants seem able to compensate for loss of leaf area; thus when sugar beet plants suffer defoliation, the remaining leaves grow larger than comparable leaves on undamaged plants, and photosynthetic efficiency is increased (French and Humphries 1977). Similar effects were reported for turnips (Taylor and Bardner 1968) but in radish neither production of new leaf area nor photosynthetic efficiency were shown to increase after experimental defoliation (Jackson 1980). Potato blight *Phytophthora infestans* can, under suitable conditions, destroy shoots within four weeks. When 75 per cent of the foliage is killed, tubers stop growing owing to the lack of photosynthates. Such devastating losses due to potato blight may not always occur; Large (1958) estimated the mean annual loss in the UK would have been less than 8 per cent over the period 1947 – 56 if main crop potatoes had not been sprayed with fungicide.

Sometimes the amount of damage is accentuated by an interaction between a pest and other insects. When field beans were infested with *Aphis fabae* 17 seeds per plant were produced compared with 56 seeds per plant in their absence, but when the aphids were attended by ants, the yield was reduced to only 8 seeds per plant (Banks and McCaulay 1967).

Plant parasitic nematodes mechanically damage roots which are unable to absorb as much water and nutrients as healthy ones, thus affecting photosynthesis in the leaves. However, economic damage usually only occurs if nematode populations have greatly increased. Translocation of water and nutrients in plants is also interrupted by fungi in the vascular tissues, e.g. wilt caused by *Verticillium* spp. on various crops.

Damage due to toxic substances

Toxic substances are sometimes injected into plants by insects when feeding or ovipositing. Miles (1968) refers to the effect of amino acids in the saliva of sucking bugs which are known to reduce the overall growth of plants on which they feed. Growth of a gall is initiated by a secretion deposited with the eggs of sawflies. The parasitic weed *Striga hermonthica* on cereals, especially sorghum and bulrush or pearl millet, possibly has a toxic or at least a growth regulatory effect on the host as well as removing water and nutrients (Parker 1978).

The deleterious effect of a higher plant on other plants by production of chemicals that escape into the environment is referred to as *allelopathy* by Rice (1974), who is convinced that such toxic interactions do occur in nature, but many of the experiments he described only suggest this but do not provide conclusive evidence. Some evidence which tends to support a toxin hypothesis has been reported by Hirano and Kira (1965), who examined the effects of exudates from the peach *Prunus persica*. It is known to be difficult to establish peach orchards on land previously used for this crop.

Damage due to spread of pathogens and other pests

In addition to mechanical transmission by contact and tools, microorganisms may be spread to host plants by insects, especially aphids, whiteflies, thrips and other leaf- and plant-hoppers. Some nematodes, fungi and weed seeds transmit certain virus diseases (Harrison 1977). Groundnut rosette virus, maize streak virus, rice tungro virus and various mosaic viruses transmitted by insect vectors are among the important causes of yield loss of food crops. Other important plant diseases, including witches' broom on cocoa and rice grassy stunt, are due to mycoplasma transmitted by leafhoppers.

Bacteria and fungi are usually carried externally on an insect and enter plants through injuries made when the insect feeds or oviposits. The fungus *Ceratocystis fimbriata* causing cocoa canker is spread by insects of the genus *Xyleborus* which make galleries in the wood, and severe damage has been caused in South America and Trinidad. *Nematospora gossypii*, a fungus causing weakening and yellowing of cotton lint, is transmitted by cotton stainers (*Dysdercus* spp., *Calidea dregei*), while a similar fungus *Nematospora coryli* affecting coffee beans is spread by *Antestiopsis* spp. Weed species are the host of some pests including insects, nematodes, fungi, bacteria and viruses (Duffus 1971). Couch grass (*Agropyron repens*) and other rhizomatous grasses including *Holcus* spp. are hosts of the fungus *Gaeumannomyces graminis* which causes take-all disease of wheat and barley.

Damage due to exudates and related problems

Infestations of aphids and whiteflies cause the build-up of honeydew, a sticky substance that encourages the growth of moulds such as sooty mould. Together, the honeydew and mould gum-up leaves, adversely affecting photosynthesis, and on some crops the quality of produce is reduced. Serious infestations of whitefly *Bemisia* are a major cause of concern in areas like the Sudan Gezira because honeydew on cotton lint interferes with spinning of the fibres.

Direct contamination of vegetables and fruit with insects, their excreta or exuviae is particularly important with prepacked, canned or frozen produce. Apart from causing a down-grading or rejection of the produce, such contamination can lead to those concerned being prosecuted for selling food unfit for human consumption. In many parts of Europe and North America, no damage is tolerated with some produce and this has resulted in cases of excessive use of pesticides, which Southwood (1979, 1981) refers to as *cosmetic spraying*. Weed seeds may also contaminate grain, and off-flavours in meat and milk can be caused by animals grazing weeds such as the wild onion *Allium vineale*.

Yield losses due to competition

The development of agricultural practices such as ploughing and crop rotation were forced on farmers principally by the need to control weeds. The extent of competition between crops and weeds depends on which weed species are present, when they emerge relative to the crop, weed density, and the rate at which the crop is established. Some crops like cotton, grow so slowly at first that the lack of early weed control can result in depressed yields. Continued use of herbicides has resulted in weeds closely related to the crop species becoming relatively more important; thus wild oats (*Avena* spp.) and black grass (*Alopecurus myosuroides*) are now major problems in the cultivation of cereal crops. Red rice is similarly important in rice-growing.

Response to damage

The ability to tolerate or compensate for damage depends on the growth characteristics of the plants which are attacked.

The effects of pests on an agricultural crop need to be evaluated carefully for a particular farming area, since differences in agronomic practices and

Table 1.1 Variation of crop loss between seasons. (Adapted from Way 1961)

Year	Loss of yield* of field beans (kg/ha)	Untreated yield (kg/ha)	Peak no. of aphids/plant
1	1681	175	3980.0
2	439	3638	34.0
3	2345	2107	1260.0
4	288	2120	85.0
5	226	3349	0.2
6	2095	0	6920.0
7	552	3500	48.0
8	2283	426	4960.0

* Difference between plots treated with an insecticide and untreated plots.

environmental factors will influence the severity of pest damage. Ideally, assessments are continued each year at a number of sites to measure seasonal effects and the impact of introducing new varieties. Way (1961) demonstrated a tenfold variation in crop loss due to aphids on field beans over an eight-year period (Table 1.1). This underlines the importance of basing pest management decisions on up-to-date information.

Considerable variation in crop loss will occur between different areas growing the same crop. The same insecticide control recommendations used on trial plots in Central Africa showed over a two fold change in yield increase (Table 1.2) between regions due to differences in climate, pest attack, soils, fertilizer and other agronomic factors. Losses in yield may not always be reflected in reduced revenue to the farmer. When the corn leaf blight epidemic occurred in the USA in the early 1970s, the financial losses were not as bad as were expected, as the market price increased for the smaller crop.

Various assessment methods can be used to try and determine the economic losses caused by different levels of pest infestation (Chiarappa

Table 1.2 Variation in yield response due to spraying insecticide on cotton in Central Africa. (From Tunstall and Matthews 1966)

Region	Year	No. of plots	Treated yield of seed cotton (kg/ha)	Untreated yield (kg/ha)	% increase
Malawi	1960	16	1285	742	73
Southern region	1961	20	1046	431	143
Malawi	1960	13	1264	448	182
Central region	1961	12	1407	648	117
Zimbabwe	1960	5	1939	724	167
Middle region	1961	5	1941	517	275

1971, 1981; Van der Graaff 1981; Walker 1981, 1983). Losses of yield need to be clearly defined in quantitative and qualitative terms that can be translated into economic ones. An agriculturalist should be consulted about the appropriate yield measurements of individual crops.

Assessment techniques

Sampling crops

The severity of pest attack within one field or locality is often irregular, so the effect of different levels of infestation should be estimated by sample surveys in which the amount of damage or final yield associated with the various levels of pest population is assessed. On perennial crops, yields may be related to the pest infestation level of the previous season. Sufficient representative samples are required to give a reliable estimate of pest incidence, crop damage and yield. Sampling may be at random but the distribution of a pest emigrating into a crop is affected by a number of factors, including the attractiveness of the crop, the location of the source of infestation, the strength and direction of the wind and the presence of hedges, trees and other obstacles around the edge of, or in, the field (Lewis 1969). Groundnut plants, for example, spaced closer together are less attractive to aphids than those spaced wider apart so the subsequent incidence of virus

Table 1.3 Comparison of sample surveys of fields to show number of samples required by different sampling methods. (Adapted from Church 1971)

District	No. of fields	Simple random sample	Stratified sample	Stratified sample with variable fraction sampled	Estimated average pest incidence (no. of fields in sample)
A	800	10	6	28	3
B	1600	20	13	28	6
C	1600	20	13	28	13
D	4000	50	32	28	16
Total	8000	100*	64+	112†	38

* 1/80 of fields sampled if average pest incidence estimated with a standard error ± 10%

+ If districts have different pest incidences, 1/133 of fields can be sampled to give same accuracy over total area

† Sampling increased to give reliable estimate for each district. If standard deviation assumed for whole area ± 80% and SE (standard error) for each district ± 15% then (80/15)2 = 28 fields to be sampled in each district.

rosette disease is reduced. Cotton stalks kept in villages for fuel often cause more severe infestations of pink bollworm in nearby fields. Similarly irrespective of field boundaries, pest attack may vary from one locality to another. The accuracy for a given amount of sampling is increased if a random sample is taken from a number of sections or strata within each field, but a reduction in the number of samples is only possible if the standard deviation within strata is less than the whole area being sampled. Equal numbers of samples are needed from each strata sampled, to estimate crop loss to the same level of accuracy in each (Table 1.3). Variation in soil fertility and other factors from one area to another will affect yields irrespective of pest infestation, so crop-loss data from this type of experiment need to be interpreted with care.

Crop-loss trials

The simplest crop-loss trials involve one plot being protected with an appropriate pesticide and the other being left untreated (Le Clerg 1971). Where losses due to weeds are being measured, the 'treated' plot is hand-weeded at frequent intervals to maintain a weed-free crop. Richardson (1981) refers to 'half-field' comparisons.

The amount of damage in the untreated plot will depend on the severity of naturally occurring pest populations. Insects in an untreated plot may move into a treated one, but fail to return, so interaction between plots must be minimized, especially where very mobile pests are involved. Plot size is discussed more fully in Chapter 7.

The difference in pest damage between treated and untreated plots may also be reduced artificially should improved plant growth attract more immigrants later than the untreated plots. Lepidopteran pests may lay more eggs on sprayed plants since the untreated ones may be covered with honeydew or exhibit other damage caused by aphids or other pests.

Where possible, large plots with a wide guard area around a relatively small sampling area are needed to get a truer assessment of crop loss. Joyce and Roberts (1959) showed that jassid and thrips infestations in unsprayed cotton fields of less than 10 feddan (4 ha) were affected by the sprays

Table 1.4 Difference between sprayed and unsprayed yields of seed cotton at Namulonge, Uganda. Untreated yield given in parenthesis. (From Reed 1972)

Year	Large-scale farm bulks (kg/ha)	Small-scale spraying trials (kg/ha)
1967	809 (355)	398 (571)
1968	489 (1144)	238 (1641)
1970	518 (1119)	198 (1618)

applied to surrounding fields in the Sudan Gezira. Reed (1972) compared results of sprayed and untreated plots of cotton in East Africa (Table 1.4) and questioned the usefulness of an untreated control in spraying trials, as the numbers of insects and their rate of increase is influenced by whether the surrounding environment is largely sprayed or unsprayed. Having treated and untreated plots further apart is not satisfactory, owing to variation in other factors such as level of pest incidence, rainfall and soil type.

Judenko (1973) described an analytical procedure for computing the losses caused by pests on crops with and without pesticides (Table 1.5).

Table 1.5 Computing the loss to silage maize from aphids in the absence of compensatory yield. (Adapted from Judenko 1973)

Total number of plants	1200
Number of plants (P) attacked by peat	300
Actual yield for unit area (Y)	2100 kg
Mean yield of unattacked plant (a)	2 kg
Mean yield of attacked plant (b)	1 kg
Actual loss $P \times (a - b) = L$	300 kg
Expected yield ($Y + L$)	2400 kg
Per cent loss = $\frac{L \times 100}{(Y + L)}$	12.5%

Cages have been used to exclude insect pests but plant growth is often adversely affected by the presence of a cage. As the timing of pest infestation in relation to plant development and maturation has a major effect on the amount of crop lost, trials should contain treatments in which protection is provided for short periods, or detailed observations of crop development made throughout the season.

George and Gair (1979) superimposed aphicide treatments (Pirimicarb) on a large number of farm sites in different areas of England and Wales. They showed that grain yield was increased by approximately 12.5 per cent if one spray was applied at the beginning of flowering when there were five

Table 1.6 Increase in cereal grain yield following one spray of pirimicarb (1974–77). (From George and Gair, 1979)

Number of aphids	Growth stage (GS) of crops*	No. of sites	Mean % increase
5–10 aphids/ear and increasing	10.5.1	8	11.5
>10–30 aphids/ear and increasing	10.5.1	8	13.9
5 or more aphids and decreasing	10.5.1	11	1.5
5 or more aphids	Later than 10.5.1	8	3.9
Never more than 5 aphids/ear		14	0.7

* As described by Zadoks, Chang and Konzak 1974.

or more aphids per ear and the number was rising. Field response was less if fewer aphids were present or the spray was applied later (Table 1.6).

While some crops have a short flowering period, flowers may open over several weeks on an indeterminate crop like cotton. Cotton flowers are 'white' for one day and can be tagged with a small label with the date of flowering, loosely attached around the peduncle. Subsequent routine inspection of the bolls will indicate which period of the flowering cycle produced bolls which matured and were harvested. In one experiment (Fig. 1.2) late flowers were lost due to moisture stress and not pest damage, indicating that control measures could be terminated.

The critical period for weed control is usually from seedling or germination during early crop growth when weeds can compete for nutrients. This critical period is longer if plant growth is slow (Table 1.7).

Table 1.7 Reduction in crop yields in Ghana (per cent). (From Carson 1978)

	Maize	Rice	Groundnuts	Cotton	Soyabean
No weed competition	0	0	0	0	0
Weed-infested for first 2 weeks	13	18	7	22	10
Weed-infested for first 4 weeks	21	—	20	37	20
Weed-infested for first 6 weeks	25	—	38	40	21
Weed-infested for first 8 weeks	40	—	38	78	24
Weed-infested for first 10 weeks	44	43	46	89	28
Weed-infested throughout	54	58	50	93	33

When the crop has established a good canopy which shades weeds, it can compete more successfully and crop losses are minimal. Kasasian and Seeyave (1969) proposed that crops like beans, tomatoes, sugar cane and yams had to be free of weeds for the first quarter to one-third of the crops' growing period.

Roberts (1976) showed that yields of onions (*Allium cepa*) are not adversely affected, provided that the weeds are not left for more than five weeks after 50 per cent crop emergence. If weeds are not controlled, almost half the applied nitrogen is lost to the weeds by the time the onions enter the period of maximum growth (Hewson and Roberts 1973).

Yield response to a pesticide treatment can also be greatly affected by crop variety. Responses obtained with some of the newer high-yielding varieties do vary considerably; an example of the differences in response to fungicide treatment of winter wheat in an area of high disease risk is shown in Table 1.8.

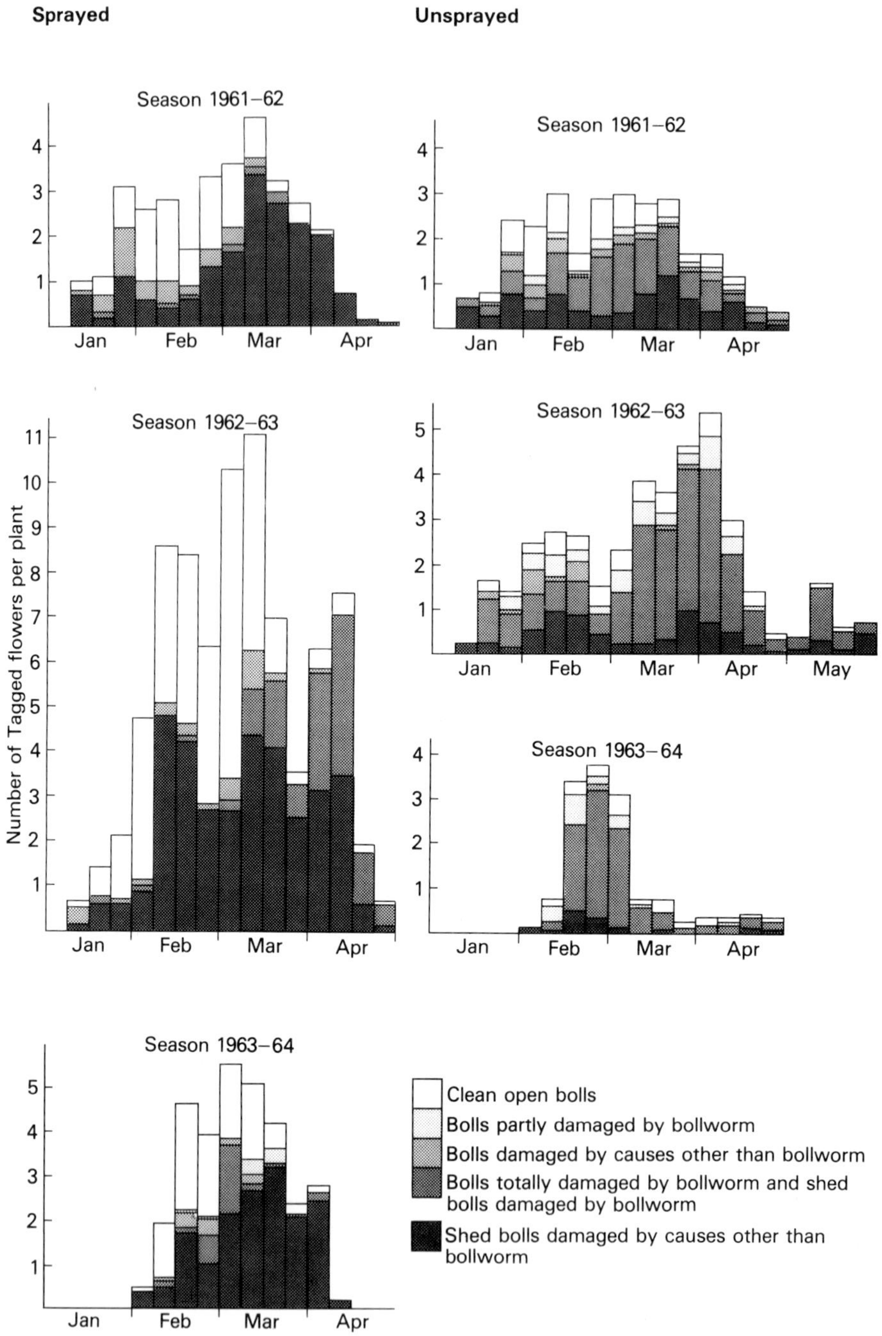

Fig. 1.2 Histograms of daily flower-tagging to show weekly totals of flowers and their subsequent fate. (a), sprayed; (b), unsprayed. (From Matthews and Tunstall 1967)

Table 1.8 Yield response to a full fungicide programme based on carbendazim, captafol and fluotrimazole in t/ha at 85 per cent dry matter. (From Clare, Boothroyd and Simkin 1981)

Year	Variety	Yield without treatment	Response
1976	Maris Huntsman 'Hobbit'	4.63	+0.44
		4.59	+0.72
1977	Maris Huntsman 'Hobbit'	7.55	−0.17
		7.11	+0.64
1978	Maris Huntsman 'Armada'	7.85	+0.37
		7.81	+0.99
1979	Maris Huntsman	6.67	+0.79
	'Armada'	6.24	+1.60
1980	Bounty	7.75	+1.18
1981	Bounty	4.98	+2.84

Simulated damage

Damage caused by pests can be simulated by removal of certain parts of plants, e.g. leaves, flowers, buds, fruit. It is difficult to simulate damage on very young buds and roots, and plants may redistribute some nutrients before a damaged part is lost, so a true assessment of actual damage may not be possible. Complete removal of buds can cause more severe damage in some cases, as an insect may only partially damage a bud and cause it to be shed later.

The effect of early season termite or grasshopper damage on cotton was simulated by removing plants at random from experimental plots (Matthews, Rowell and Beeden 1972). During the remainder of the season plants were protected from bollworm damage by routine spraying. Climatic conditions allowed the branches to continue to grow and occupy the inter-row space, so in plots with few plants each plant had more bolls (Fig. 1.3). Individual bolls weighed the same irrespective of plant density, except where a lack of soil nutrients produced smaller bolls on plots having the most plants. The maximum yield was obtained with 3–4 plants/m^2 but the yield was also high when only 10 per cent of the original plant population remained (Table 1.9).

Table 1.9 Average yields of fertilized cotton following simulated early season pest attack. (From Matthews, Rowell and Beedon 1972)

% stand	Number of plants/m^2	Yield of seed cotton (kg/ha)
10	0.7	1841
100	7.2	2145

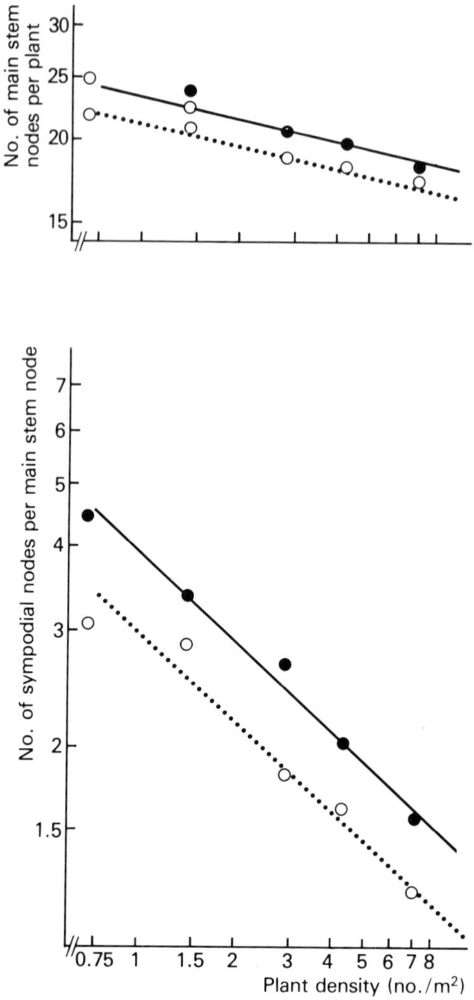

Fig. 1.3 Effect of loss of plants on yield components. (From Matthews, Rowell and Beeden 1972)

Where there were few plants, bolls were exposed to bollworm attack over a much longer period, which suggests that poor stands would be undesirable in a pest management programme.

Artificial infestations

Assessment of damage is very difficult if the natural pest infestation remains low, so damage can be increased by artificially infesting plants. This is particularly useful when assessing yield loss on susceptible and tolerant varieties, provided sufficient numbers of a pest can be obtained from a

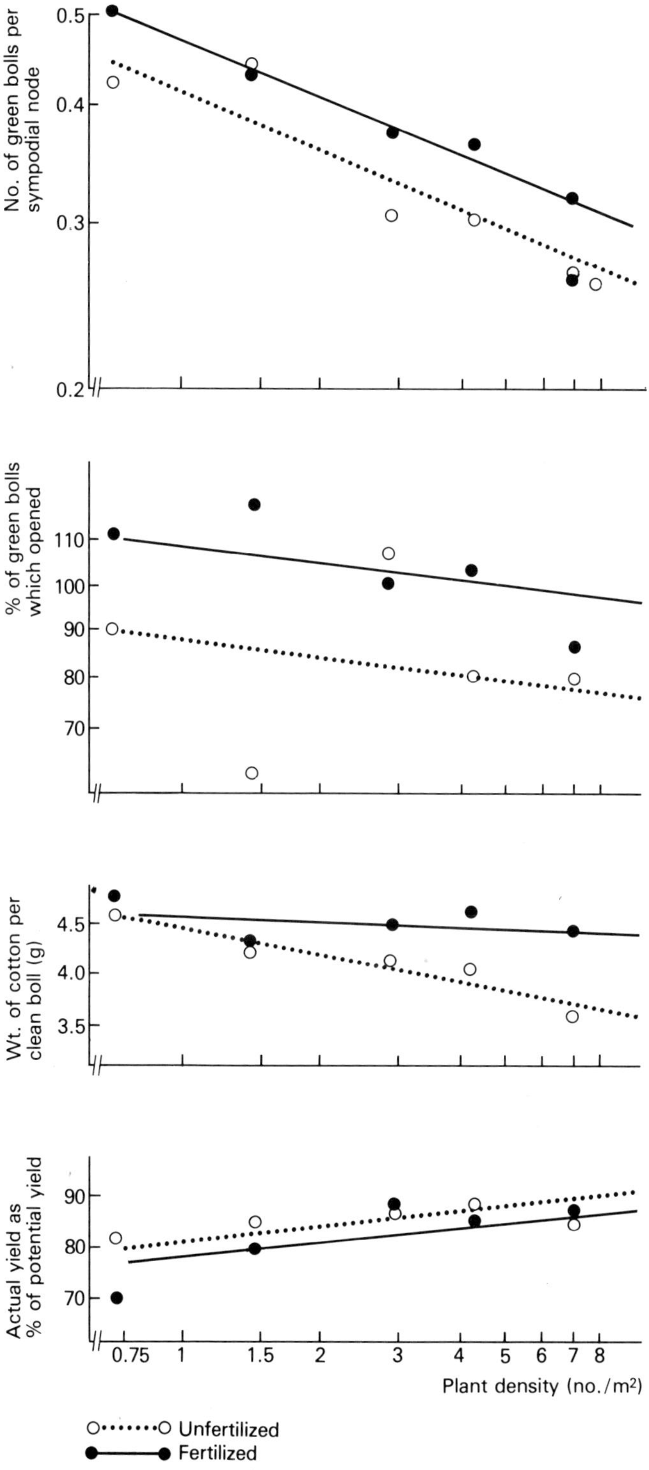
No. of green bolls per sympodial node
0.5
0.4
0.3
0.2
% of green bolls which opened
110
100
90
80
70
Wt. of cotton per clean boll (g)
4.5
4.0
3.5
Actual yield as % of potential yield
90
80
70
0.75
1
1.5
2
3
4
5
6
7
8
Plant density (no./m²)
Unfertilized
Fertilized

laboratory culture or collected from the field. Similarly, inoculating seed or plants with different spore loads allows the effects of diseases on yields to be investigated; or plots can be sown with different numbers of weed seeds.

There is little point in introducing an artificial infestation unless conditions are favourable to maintain it. Where high humidity is needed plots can be covered within a transparent plastic tunnel or individual bags (Russell 1978). This system has been shown to provide suitable conditions for the development of powdery scab on potatoes (Jellis 1974). Irrigation and misting of trial plots has also been used to promote disease.

Such experiments with insects often involve caging plants or parts of plants to prevent the introduced pest from emigrating, and also to exclude predators which might otherwise be attracted to the artificial infestations. The relative impact of different numbers of a pest can only be fairly assessed if any adverse effects due to the presence of the cage are similar for all sample populations, including those without pests. A non-persistent pesticide such as a natural pyrethrin may be needed to clear foliage of insects before the cages are first put in place. The effect of different times of infestation on crop loss can also be studied.

Evaluation of crop losses

Tammes (1961) and Bardner and Fletcher (1974) described a generalized response curve (Fig. 1.4) to indicate the relationship between intensity of

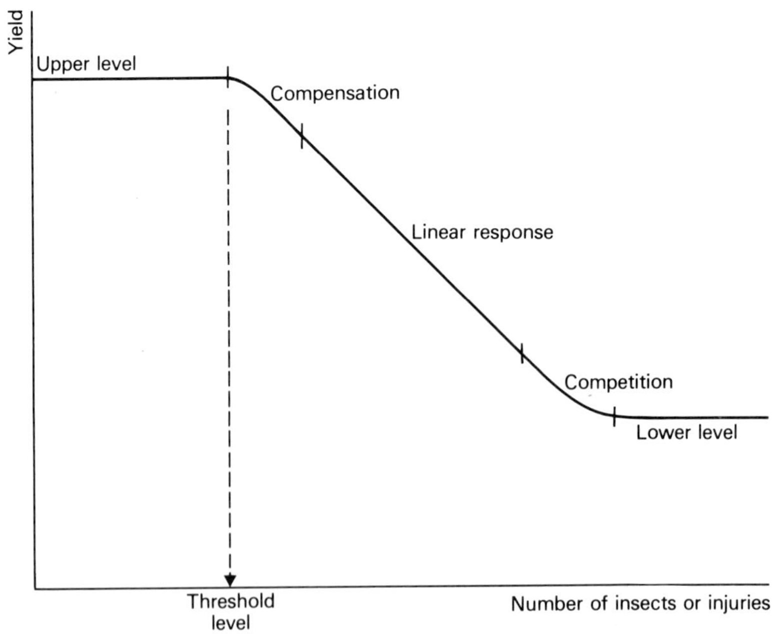

Fig. 1.4 Generalized response curve. (From Bardner and Fletcher 1974)

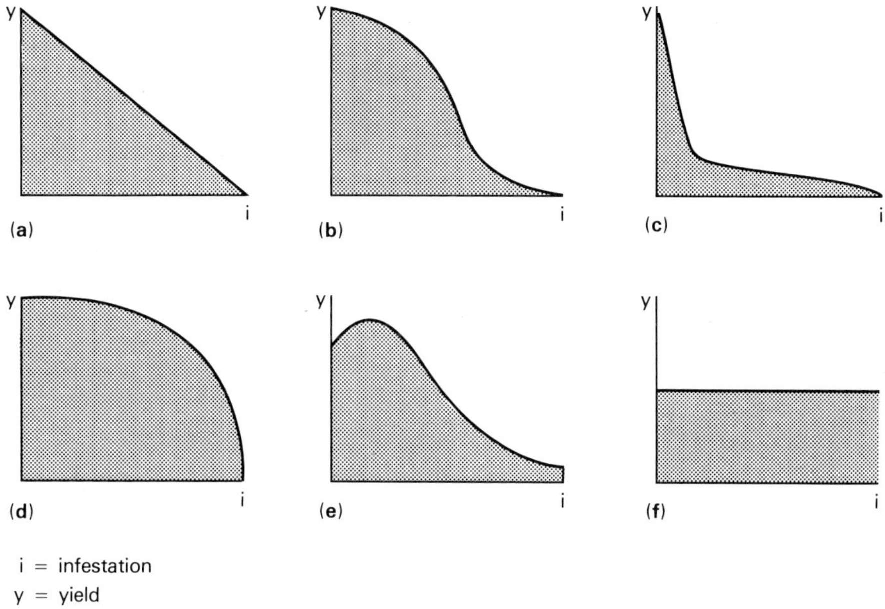

Fig. 1.5 Variations in yield response to infestation. (From Walker 1977)

infestation and yield, but Walker (1977) has gone further and proposed six types of relationship (Fig. 1.5). In a few cases yields may be reduced proportionally as infestation increases (a) but a sigmoid response is most common (b). Severe damage due to a very low pest population may occur when a few vectors transmit a virus or other disease-causing organisms (c). Yield may also be logarithmically related to the number of pests (d), or increase with a light infestation (e). The latter may occur when a light infestation of *Heliothis* stimulates bud production. In some experiments an increase in infestation has produced no significant change in yield, possibly because adequate compensation by the plant is possible (f).

When the extent of crop losses have been determined, the value of such losses needs to be assessed and related to the control costs. Southwood and Norton (1973) illustrate how the cost–pest and pest–revenue relationships can be combined (Fig. 1.6). The response to an insecticide aimed at control of a foliar pest is likely to be high initially but will soon level off. Kogan (1976) showed that a pest such as the green cloverworm on soyabean can be controlled adequately by one carefully timed low-dosage application, so neither increasing the dosage nor the number of sprays will decrease the level of injury further.

When the harvested produce is damaged directly, the initial revenue response may be low, but additional control measures will result in more revenue. On some crops, the increase in revenue is mainly associated with quality of the product. This has led to the use of cosmetic sprays to obtain blemish-free produce. Monitoring pest numbers, discussed more fully in the next chapter, is needed to avoid excessive applications of pesticide, and is

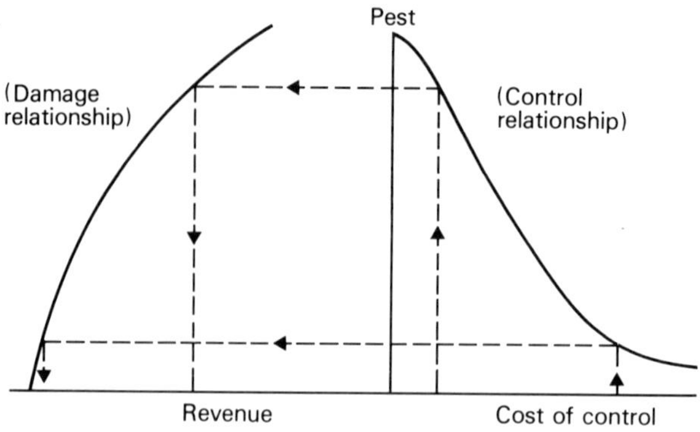

(a)

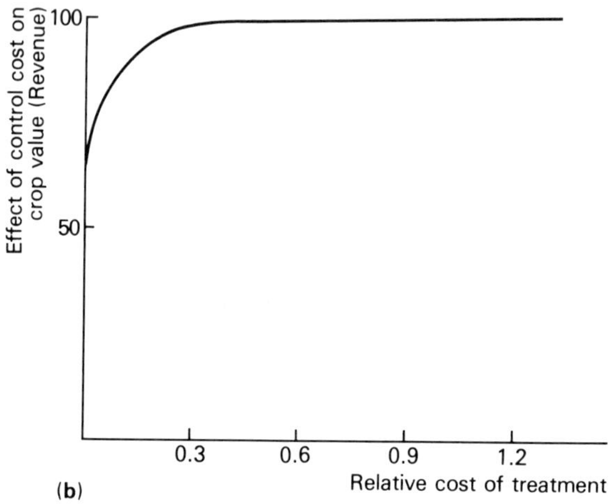

(b)

Fig. 1.6 (a) Combination of control and damage function
(b) Effect on revenue with a foliage-attacking pest. (Adapted from Southwood and Norton 1973)

likely to result in a more economic control programme. The improved revenue using an optimal spraying strategy was illustrated by Southwood and Norton (1973) using data from cotton trials (Fig. 1.7). Successful implementation of pesticide applications timed in relation to economic thresholds depends on having sufficient equipment to treat areas quickly enough.

In some crops the frequency with which different levels of damage may occur within particular areas is the important factor so that decisions can be made on the probability that control measures will be economic. Rijsdijk and Zadoks (1979) have suggested preparing maps of 'iso-loss' and 'iso-risk' to present such data.

Attempts to control pests in North America, Europe and Japan over the last few decades have relied very much on the use of pesticides, but the

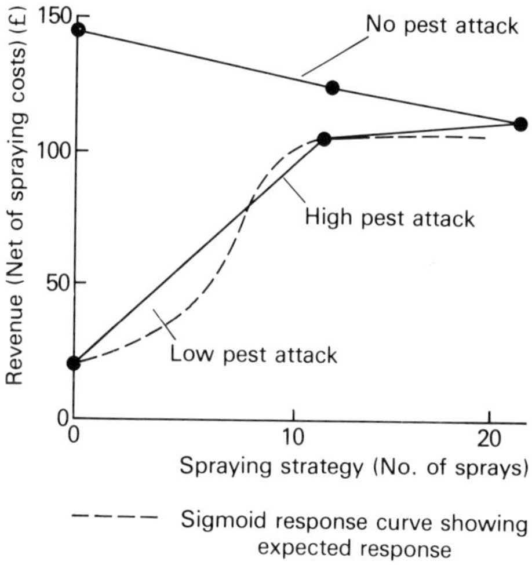

Fig. 1.7 Potential variation in revenue with different spraying strategies based on data for cotton – Matthews and Tunstall, 1968. (From Southwood and Norton 1973)

emphasis is now on the development of pest management programmes integrating different control techniques. So far this has depended to a large extent on experiments (Way 1973), but in an attempt to categorize pests and provide a theoretical approach to integrated control, Conway (1976) has classified pests in relation to the spectrum of ecological strategies described by Southwood (1976), and has given examples of insect pests which can be regarded as *r* or *K* pests (Table 1.10). At one end of the spectrum *r* pests have a high potential rate of reproduction, so populations can increase very rapidly as soon as environmental conditions – food supply, temperature, humidity and other factors – are favourable. Populations can rebound rapidly even when natural enemies or disease cause heavy mortality. Such pests, e.g. aphids, are readily dispersed and are successful at finding and exploring new habitats, resulting in massive outbreaks which can cause extensive damage before the natural enemies can act. Migrant pests such as locusts, some armyworms and birds such as quelea are *r* pests. Many weed species are *r* pests and exploit habitats created by man and interfere with agricultural activities. In contrast *K* pests have a relatively slow rate of increase and have a more stable population. The competitive ability of *K* pests is greater, but they prefer more specialized food. Low populations are regarded as important; thus codling moth is a major pest of orchards as there is a demand for high-quality, unblemished apples. Similarly only a few tsetse flies in an area enable trypanosomes to be transmitted.

Important pests are often either *r* or *K* strategists, but most pests are intermediate between these extremes. Populations of intermediate pests are normally regulated by natural enemies, pest outbreaks occurring when this process is interrupted. Conway (1976, 1977) suggested that the principal

Table 1.10 Some important insect pests and their characteristics. (After Conway 1976)

Species	Fecundity per females (approx)	Generation time (approx)	Pest features
r pests			
Desert locust (*Schistocerca gregaria*)	400	1–2 months	Periodic outbreaks which defoliate any crop
Black bean aphid (*Aphis fabae*)	100	1–2 weeks	Wide range of crop plants including beans and beet
Housefly (*Musca domestica*)	500	2–3 weeks	Feeds on organic matter
Black cutworm (*Agrotis ipsilon*)	1500	1–1.5 months	Attacks seedlings of most crops
K pests			
Rhinoceros bettle (*Oryctes rhinoceros*)	50	3–4 months	Adult feeds on apical growing point of coconuts often killing pod
Tsetse fly (*Glossina* spp.)	10	2–3 months	Vector of trypanosomiasis affecting man and cattle
Codling moth (*Cydia pomonella*)	40	2–6 months	Larvae damage apples and other fruit
Sheep ked (*Melophagus ovirus*)	15	1–2 months	External parasite of sheep

control technique appropriate for *r* pests was the use of pesticides, as this was the only method with speed and flexibility required to respond to an outbreak of the pests. Pesticides are also appropriate for *K* pests when very small populations cause considerable losses of revenue, but preference should be given whenever possible to alternative strategies such as cultural control and use of resistant varieties or animals (Fig. 1.8).

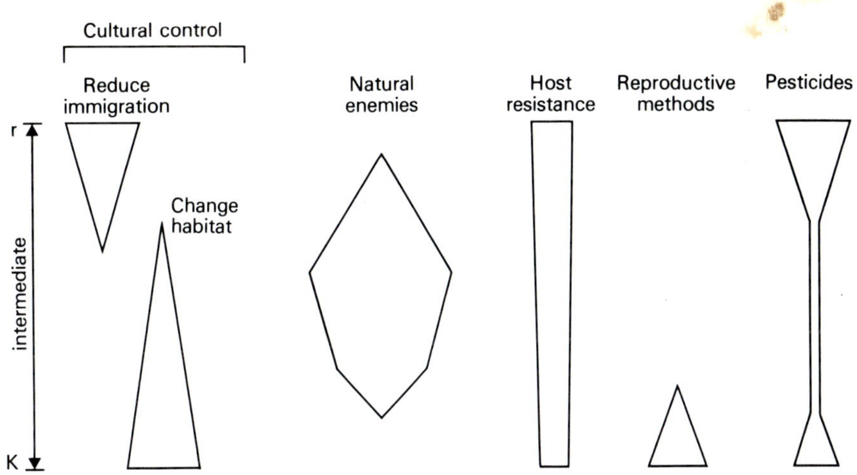

Fig. 1.8 Appropriate control strategies for pests categorized in terms of the *r–K* continuum. (From Conway 1977)

2

How many pests?

The number of pests in a crop may remain so low that no economic damage is caused, but when a pest population density has exceeded the economic injury level (EIL), control measures must be implemented to avoid loss of revenue. Clearly the value of the crop saved must exceed the cost of treatment. Action at this stage may involve the application of a pesticide to check rapidly any further increase in the pest population; the choice of actual control technique is discussed more fully in Chapter 3. The EIL is difficult to determine: final yields cannot be forecast when so many environmental factors influence not only subsequent crop growth but also the severity and duration of a pest attack. Lack of knowledge about economic injury levels has encouraged many farmers to play safe and initiate controls, even when the pest is not present. Prophylactic treatment may be justified if crop damage is significant in most seasons when the cost of treatment is relatively low, and the effects can be assessed over several years. When a low pest population can spread a virus prophylactic treatment may be valuable, since the initial infection of the vector may be very difficult to detect. A single viruliferous green aphid *Myzus persicae* found per four plants in a sugar beet crop has been considered to be a dangerous population (Heathcote 1978). As sugar beet yields can be severely reduced by virus yellows in some seasons, many growers have preferred to apply a granular insecticide such as aldicarb at planting, to protect young seedlings rather than wait to detect an aphid infestation and then spray. However, continuous exposure to insecticide residues in a crop will encourage selection of pesticide-resistant aphids, so early warning systems are preferred to avoid unnecessary use of pesticides in years of low pest infestation.

Ideally, the farmer or his advisers need to monitor changes in pest populations regularly throughout a season so that decisions can be made quickly. Experience of population dynamics within particular crop localities will increase the accuracy of such decisions and eventually limit sampling to crucial periods. The threshold for some pests may occur more or less at the same time each year, though the exact dates will often depend on differences in temperature or some other seasonal factor.

The methods of assessing pest populations given in the following sections have been selected to represent those most widely used in pest management programmes. Walker (1981) also lists methods of measuring insect populations. Other techniques used in ecological studies are given by Southwood (1966) who points out that methods to measure the relative changes in populations are quicker than attempts to determine absolute numbers. The results of relative methods will, of course, depend on the actual density or population size, numbers in a particular phase or level of activity, efficiency of sampling method and responsiveness of a particular sex and species to the traps being used, but changes in population detected can be used to assess when control measures are needed.

Techniques for estimating insect pest populations

Crop inspection

Regular inspections by walking through crops is essential, not only to examine the status of insect pests, diseases and weeds, but also to assess plant development. Monitoring pest levels in this way is often referred to as *scouting* and has been adopted on a large scale on cotton, particularly in some areas of the USA such as Arkansas (Lincoln *et al*. 1970) and in Central Africa (Burgess 1983). Scouting is particularly important in areas where the severity of pest attack may vary considerably from one year to another or where the period of infestation varies from year to year. Trained scouts may be resident on a farm throughout the season to collect data for the farmer's own interpretation, though in some countries vacation students or consultants are employed to visit fields regularly and advise on appropriate action.

On most crops sampling of eggs or early larval instars is recommended where pest populations can build-up rapidly, and action has to be taken before serious damage occurs. Bollworm egg counts were recommended on cotton by Matthews and Tunstall (1968), but more recently preference has been given to larval counts early in the season before insecticides are applied. The aim is to avoid use of an insecticide too early in the season when important egg parasites may be killed. Also, a virus can check the population. Scouting individual fields is economic when pest infestations occur over a prolonged period, but for many crop-pest situations routine sampling of fields requires too much skilled or semi-skilled labour, and alternative systems are needed.

Accuracy of egg counts will depend on the eyesight, particularly colour vision, of the scout. However, larvae may be generally difficult to detect, especially if they are inside buds or stems. Another disadvantage with certain

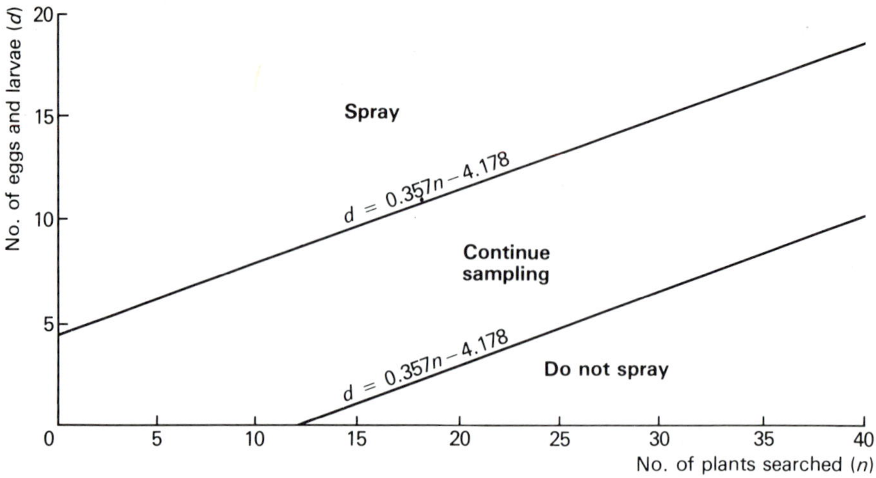

Fig. 2.1 Sequential sampling graph for use when counting eggs and larvae. (From Ingram and Green 1972)

crops is the time involved in searching a sufficiently large sample accurately. The size of the sample can be reduced by using a sequential sampling, but to devise such a system basic data to provide a reliable estimate of pest populations must be available (Southwood 1966; Ruesink and Kogan 1975; Sterling and Peters 1979). In the example (Fig. 2.1) the decision to spray can be taken after only a few plants are found heavily infested, whereas sampling can be stopped and no spray applied if thirteen plants have no pests. Alternatively, selected parts of plants may be inspected instead of whole plants; thus the top half of cotton plants where *Heliothis* eggs are laid has to be examined. Some pest species can be found easily when the crop is inspected during daylight, although many insects are more active at night, and a visit to fields after sundown is necessary to observe pest behaviour within the crop. A torch fitted to a cap (or a miner's lamp) allows both hands to be free to use a net or examine plants.

Assessment of the crop itself is important in order to relate pest numbers to the growth stage of the crop, particularly where further plant development will compensate for the expected damage caused. In some cases the numbers of beneficial species can also be estimated; thus Croft (1975) working on the control of apple pests recommended a decision index which related numbers of predatory mites and spider mites to the amount of leaf bronzing. Thus an acaricide is only applied if there are insufficient predators to cope with the pest population (Fig. 2.2). Ooi (1982) has developed a surveillance system for rice in Malaysia in which predator populations are monitored along with planthoppers while Room (1979) reported a pest management system on cotton in Australia which involved daily sampling of ten pests, twenty predators and six parasites.

Counting large numbers of a small pest such as mites is difficult and time-consuming (see page 26), but rapid assessments can be made in the field with a scoring method which grades infestations. Over 30 adult female spider

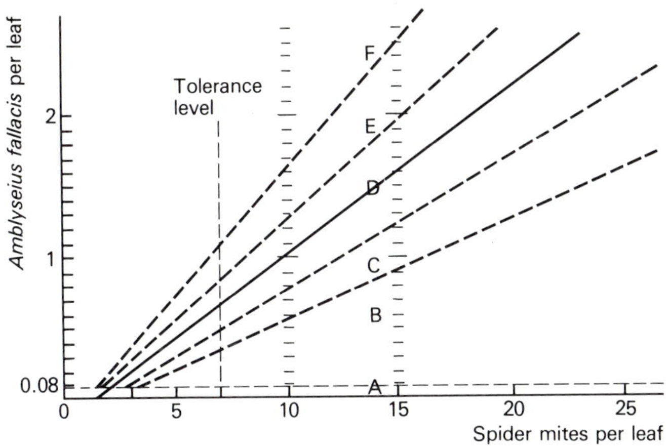

Region	Suggested recommendation	Probability for biological control
A	As bronzing appears, spray recommended miticide at full rate	Very low
B	If bronzing appears, spray Plictran† 50 w.p. at 2oz/100 gal or 8 oz per acre	Equal to or less than 10%
C	If bronzing appears, spray Omite‡ 30 w.p. at 1 ¼ lb/100 gal or 5 lb per acre or Plictran 50 w.p. at 1 ½ oz/100 gal or 6 oz per acre	Greater than 10% but less than 50%
D	Wait 1 week; biological control should occur soon; if not, spray Omite at 1 ¼ lb or Plictran 1 ½ oz/100 gal	Approximately 50%
E	Same as D	Greater than 50% but less than 90%
F	Wait one week; biological control is almost certain	Greater than 90%

† Trade name for tricyclohexylhydroxytin
‡ Trade name for 2-(*p-tert*-butylphenoxy)cyclohexyl 2-propynyl sulphite

Fig. 2.2 Decision-making index for estimating biological control of spider mites by *Amblyseius fallacis*. (From Croft 1975a)

mites per plant can be scored as 3, 11–30 as 2, 1–10 as 1 and 0 as 0. Three leaves, one leaf from the middle region and the others from the top, are inspected on each plant. Generally, when this system is used economic damage occurs if, on cotton, an average score of 2 per plant is exceeded before first boll split, but later heavy infestations causing defoliation make harvesting easier (Duncombe 1977).

Scouting a crop can also involve one or more of the trapping methods described below.

Shaking and beating

Some insects may be dislodged from plants by shaking. A sheet of cloth or polythene is normally laid out under the plants, parts of which are then vigorously shaken or beaten with a wooden stick. The insects need to be collected quickly before they escape, so a pooter (a small device for sucking insects into a tube) is useful. Cards with sticky surfaces have also been used, especially where there is sufficient space around the plants (e.g. with small bushes and row crops). However, the numbers collected may not be correlated with those from other sampling techniques, since the amounts of foliage and intensity of shaking will affect the proportion of insects dislodged. Thus, numbers of bean leaf beetles on soya obtained by beating and sweep net samples could not be correlated (Kogan, Ruesink and McDowell 1974).

'Knock down' sampling

An irritant spray is a useful alternative to shaking especially when large trees have to be sampled. Natural pyrethrin or more recently non-residual synthetic pyrethroids, synergized with piperonyl butoxide, have been applied with a knapsack mistblower to project the spray as high as possible into a tree canopy (Southwood, Brown and Reader 1979). A large area under the sampled tree needs to be covered as some insects knocked down by the spray may be blown downwind. Sufficient time must be allowed for all the insects affected by the spray to fall. Knock down samples have been used on coffee in East Africa for many years to determine when to control Antestia bug with insecticides (Wheatley 1962). When sampling mosquitoes in houses a small hand-operated flit gun sprayer can be used.

Bait sprays

Molasses added to an insecticide such as carbaryl can be sprayed along a row crop to sample moths, but the catch needs to be examined very early the next day before ants have removed the bodies.

Mite sampling

A subjective assessment of mite populations is often the only means of coping with a large number of samples, but if more detailed counts are needed, particularly of adult mites, it is possible to brush them off leaves on to a glass plate for counting. Small-mite brushing equipment has been described (Henderson and McBurnie 1943; Chant and Muir 1955; Chant 1962). Leaves are dipped between contra-rotating spiral brushes which dislodge the mites. These then fall on to a rotating glass plate coated with a very thin layer of adhesive such as polyoxyethylene sorbitan monolaurate (Tween 20) (Cleveland, 1962) or varnish. Counting may be with the aid of a special grid or disc (Strickland 1954). It is important that the brushes be set carefully so as to remove the mites without crushing them.

An alternative sampling method is to place the leaf on a clean piece of absorbent paper and to squash the mites with a roller. The stains on the paper are then counted; and the paper can be kept for checking.

Sweep net catches

A relatively inexpensive means of sampling some insects such as planthoppers is by sweeping briskly with a 40 cm diameter net made from white linen or calico cloth fitted to a short handle. The net is swung horizontally backwards and forwards through an arc of 180° with the open end perpendicular to the ground. On rice, ten sweeps per plot are recommended, particularly for leafhoppers feeding on the upper parts of plants (the top 25 cm), with one sweep for each step forward (Heinrichs *et al.* 1982). The mouth of the net must be closed as soon as sweeping is completed. The efficiency of sweeping does vary with the methods adopted by different individuals, the stage of the crop – and particularly its height, weather conditions including wind speed, temperature and intensity of solar radiation, and the diurnal rhythm of the insect. Care must be taken to avoid the escape of fast-flying insects as soon as the net is opened to examine the catch. Sweeping is not suitable for pests normally found near the base of plants such as planthoppers on rice since populations are thus liable to be underestimated.

Malaise trap

Malaise traps are best used for certain highly active adult Diptera and Hymenoptera, and so are useful in assessing the abundance of certain parasites within crops. The trap consists of a netting tent with one open side into which is placed a small container at the highest point (Fig. 2.3). Insects which fly or crawl through the opening normally move up the netting and are collected in the container. The traps are unsuitable for insects as small as aphids because the screen mesh needs to be so fine that shelter effects occur and airflow through the trap is interrupted.

Impaction traps

Insects can be trapped on a surface which is coated with a substance that remains sticky over a long period. The catch of certain species may be increased if an attractant is added to the sticky material or the surface is of a particular colour. Sticky traps for sampling aphids (Broadbent *et al.* 1948) can be made from a 30 cm length of yellow plastic wrapped around a 12 cm

Fig. 2.3 Malaise trap

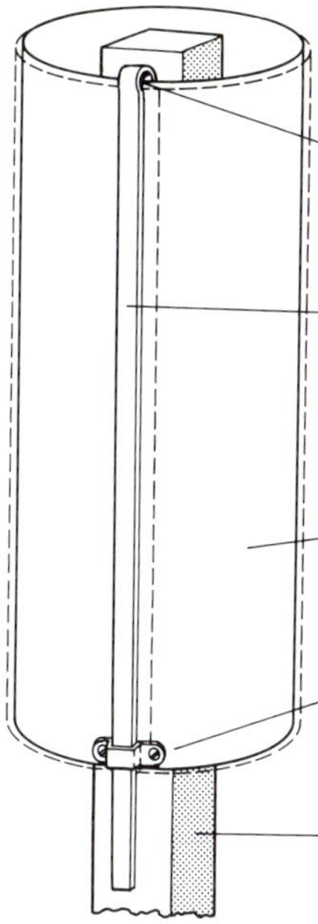

Fig. 2.4 Cylindrical sticky trap. (After Broadbent *et al.* 1948; reproduced from Southwood 1966)

diameter pipe (Fig. 2.4). Banding grease is smeared on the plastic cover and the cylinder supported on a wooden stake at a height of about 1–2 m. Efficiency of these traps was found to be almost constant with wind speed from 1 to 5 m/s, so they were chosen by Taylor (1962a) in preference to water traps.

Trap catches are affected by rain and other weather conditions, and the position and height of the trap in the field. Flat sticky plates are generally not as effective as cylinders but have been tried mounted horizontally to indicate the number of insects alighting on a surface. Flat sticky boards have also been used to sample rice pests. Sticky bands can be placed around tree trunks to sample pests moving up from the ground. In this way infestations of grasshopper nymphs are monitored on *Pinus patula* in Malawi, to determine which parts of a forest were liable to defoliation.

Water traps

Yellow water traps have been used by various workers to sample aphids. Various containers have been used to hold water plus detergent and a preservative, but Taylor and Palmer (1972) propose a circular trap coloured to British Standard Prime Yellow (BS 0.001). Aphids can be caught at low density with these traps, even when they are not with a sticky trap. Water traps are usually more effective if raised above ground level, but height will

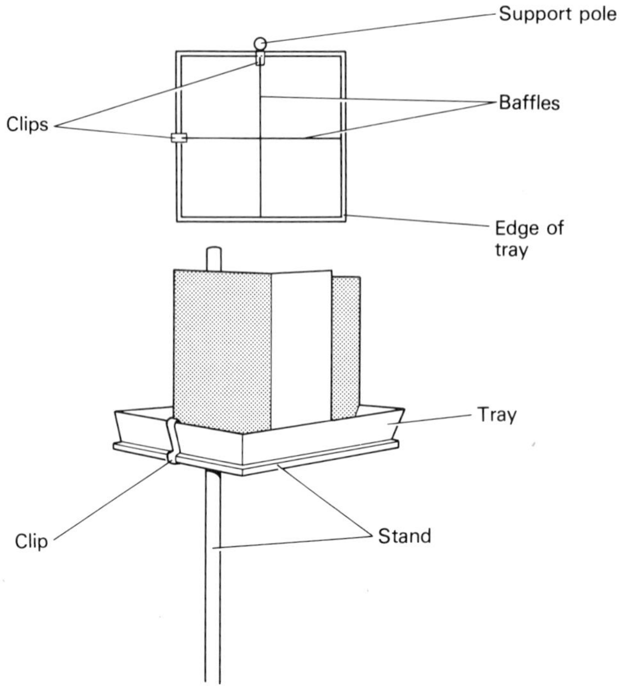

Fig. 2.5 Water trap with baffle. (From Southwood 1966)

depend on the species being sampled and the time of the year. For example, more green peach aphids *Myzus persicae* were collected in traps up to 4.75 m from the ground in late summer and autumn when the aphids migrated to peach trees (Landis 1972). Vertical baffles placed in the trap can increase the catch (Fig. 2.5). Simple yellow plastic buckets have been used to sample rice pests, and do not require so much attention as shallow traps which are liable to overflow if heavy rain occurs, or dry out in the sun. Catches of more than fifty brown planthoppers, mostly males, should be evidence of an important immigration liable to cause severe hopperburn (Kisimoto 1979) but Heinrichs (personal communication) has not found yellow water traps useful for timing insecticide treatments, preferring kerosene light traps where electricity is not available. In India, water traps have been used in conjunction with fishmeal bait to attract sorghum shoot flies.

Suction traps

Mobile traps

Various machines have been developed with engine-driven fans to suck insects into a fine mesh net. A knapsack version with a 2-stroke 98 cc engine and a centrifugal fan described by Dietrick (1961) and Thornhill (1978) has a large opening up to 35 cm diameter which can be placed over plants

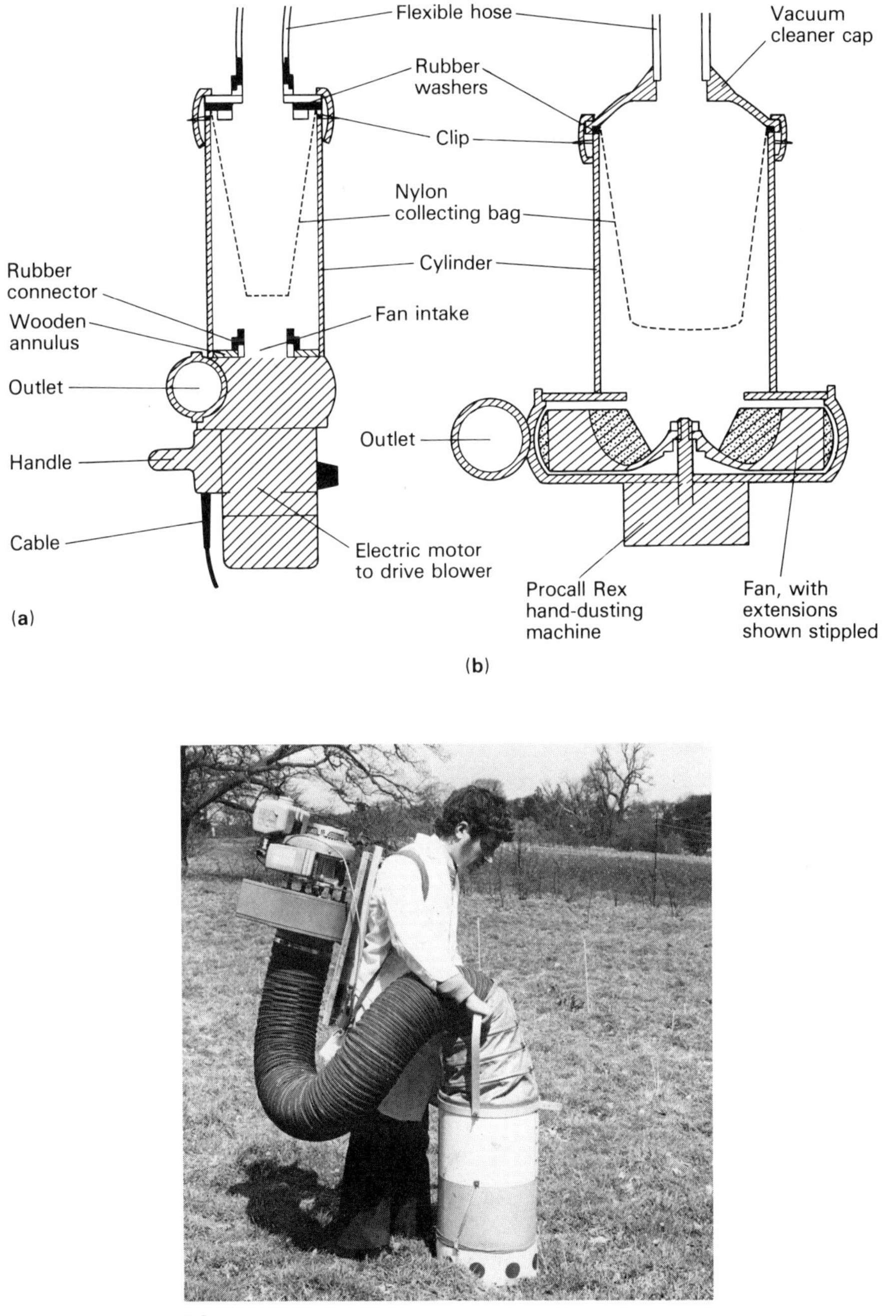

Fig. 2.6 Suction apparatus. (From Southwood 1966)
(a) Electric model. (After Johnson, Southwood and Entwistle 1957)
(b) Hand-operated model. (After Southwood and Pleasance 1962)
(c) Engine-driven model

(Fig. 2.6c). These machines are heavy, noisy and cumbersome to operate for long periods, and may not always effectively sample all the insects, particularly some of the Heteroptera nymphs in a given area. The collecting bag is positioned at the opening of the long hose so damp foliage can be sampled without insects being trapped in a water film in the hose, though sampling should be avoided in wet weather.

A smaller collecting hose on similar machines (Arnold, Needham and Stevenson 1973) can be used to probe and sample particular sections of a crop canopy or the ground demarcated by a frame. An air velocity at the nozzle of at least 25 m/s is needed if high extraction rates are to be obtained. A motorized knapsack mistblower can be adapted by fitting a hose to the suction side of the fan. Small suction traps may have an electric engine (Johnson, Southwood and Eutwistle 1957) powered by a portable generator. The smallest of these samplers has batteries as the power source (Cinch 1971; Carino, Kenmore and Dyck 1979). Sampling with these traps is very useful in assessing the effects of pesticides on the populations of parasites and predators.

Fixed traps

Small suction traps for catching insects in crops have been described by Johnson (1950) and Taylor (1951). The exposed cone type (Fig. 2.7a, c) usually has a 30 cm fan through which air and insects pass into a copper gauge cone at the end of which is a collecting tube. Smaller fans can be used only in very sheltered sites, as cross-winds in excess of 4 m/s significantly reduce the air intake of the fan. The catch can be separated at predetermined time intervals by a disc mechanism so that the period over which flight occurs can be studied. Other traps with an enclosed cone have been designed so that insects do not pass through the fan (Fig. 2.7b)

In recognition that pest control required a national, if not international, scale of operations (Williams 1952), routine sampling of airborne insects is maintained in the UK with a network of twenty-one standardized suction traps to survey migrant species and aphids (Taylor 1979). Early warning of the likely arrival of immigrants in crops is possible at distances up to 40 km from a trap (Taylor and Palmer 1972). The trapping system has been extended in to western Europe. In 1979 alates of the aphid *Metopolophium dirhodum* were first detected in the south of France, with a gradual spread northwards so that the population peak in East Anglia (Fig. 2.8a, b) occurred in late July (Dewar, Woiwod and Choppin de Janvry 1980). The increase in population was correlated with latitude and probably also with crop development.

Large suction traps were developed to obtain a sufficiently large sample of air and estimate the aerial density of insects (Johnson and Taylor 1955). The number of insects caught is divided by the volume of air sampled so an accurate assessment of air volume is needed to correct the data for efficiency of extraction (Taylor 1955). Size of insect and wind speeds affect the correction factor (Taylor 1962b). The 12.2-m high Rothamsted suction trap (Fig. 2.9) has a centrifugal fan inside a box of 76 cm × 3 m at the base of a 9 m plastic chimney, 25 cm diameter. A large wide-mouthed cone and the expansion chamber slow down the air speed, which can reach 56 km/h with large volumes of air (> 2500 m^3) being sampled. Trap height was chosen to increase the proportion of aphids in the sample and thus reduce the cost of sorting catches. Sorting is laborious and requires trained manpower to differentiate between species, and Lewis (1981) has discussed the problems in collecting and processing data more rapidly so that information is available in time to improve short-term decisions in pest management. Also the usefulness of a survey system depends on knowing the relationship between trap catches and the immigrant density on crops. This has been achieved so far only for *Aphis fabae* (Way *et al.* 1981). Suction traps undoubtedly provide the best independent sampling method for measuring absolute population density in nearly all weather conditions. Suction trap catches are more consistent than those of sticky traps, but both types were shown to be better than crop inspection at detection of the first seasonal immigration of *Myzus persicae* and *Aphis fabae* into sugar beet (Heathcote, Palmer and Taylor 1969). Apart from knowing when these immigrants arrive, it is important to know whether they are viruliferous. Enzyme-linked immunosorbent assay (ELISA) and immuno-specific electron microscopy techniques can detect virus in freshly killed aphids and may be applicable to preserved specimens (Lewis 1981).

Light traps

Light traps (Fig. 2.10) have been widely used for trapping insects. Radiant energy in the near ultra-violet region (320–400 nm) is very attractive to a wide range of insects, so most light traps now use an ultra-violet lamp, although kerosene pressure lamps have been widely used. The lamp is usually surrounded by baffles above a funnel down which insects fall into a container. Rapid kill of the catch is essential to have good specimens for identification. Pieces of plastic strip impregnated with dichlorvos have been used and are very effective in the confined space of the trap. Frost (1952) Hollingsworth, Harstock and Stanley (1963) and Southwood (1966) refer to various designs of trap, including the Rothamsted trap, originally fitted with a 200 W tungsten lamp (Williams 1948) and the Robinson trap (Robinson and Robinson 1950), the first to use ultra-violet light.

The Rothamsted trap has a 60 cm square roof over the lamp confining the illumination to a beam which diverges about 22.5 ° above the horizontal in all directions. The Robinson trap lamp is exposed to a complete hemisphere. This trap was modified for use in East Africa by Brown (1965), who replaced the plastic cone with a metal cone and protected the lamp with an inverted 2 litre glass beaker. The modified trap is known as the Muguga trap. Catches with Muguga mercury vapour trap were fifty times as big as those from a standard Rothamsted trap (Taylor and Brown 1972).

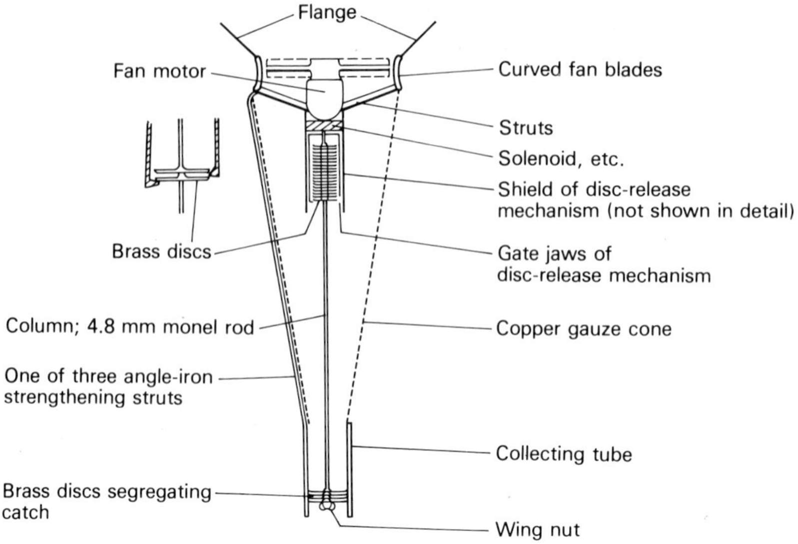
Flange
Fan motor
Curved fan blades
Struts
Solenoid, etc.
Shield of disc-release mechanism (not shown in detail)
Brass discs
Gate jaws of disc-release mechanism
Column; 4.8 mm monel rod
Copper gauze cone
One of three angle-iron strengthening struts
Collecting tube
Brass discs segregating catch
Wing nut

(**a**)

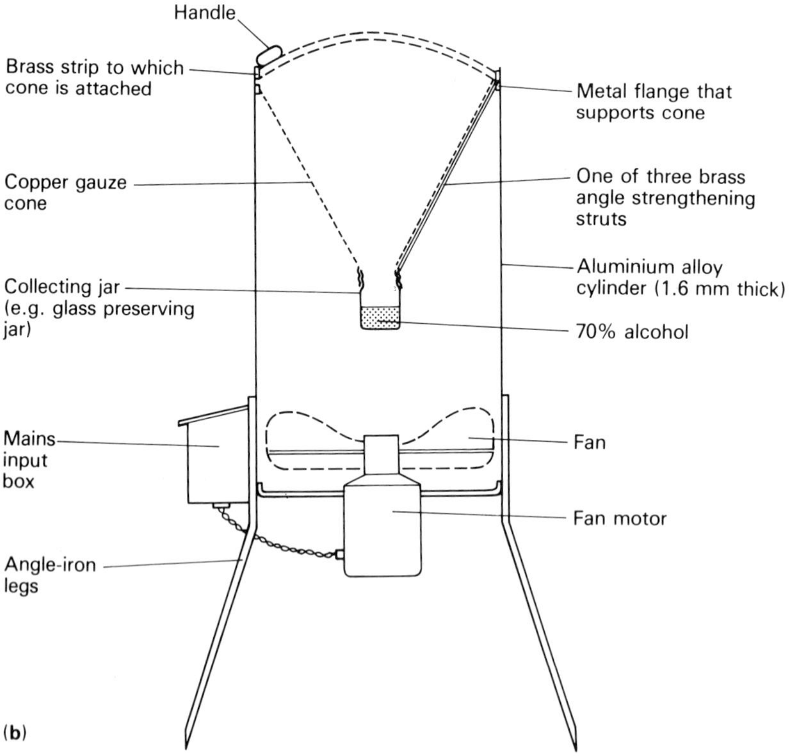
Handle
Brass strip to which cone is attached
Metal flange that supports cone
Copper gauze cone
One of three brass angle strengthening struts
Aluminium alloy cylinder (1.6 mm thick)
Collecting jar (e.g. glass preserving jar)
70% alcohol
Mains input box
Fan
Fan motor
Angle-iron legs

(**b**)

(c)

Fig. 2.7 Suction traps
(a) Exposed cone type. (From Southwood 1966)
(b) Enclosed cone type (From Southwood, 1966)
(c) Exposed cone trap in cereal field

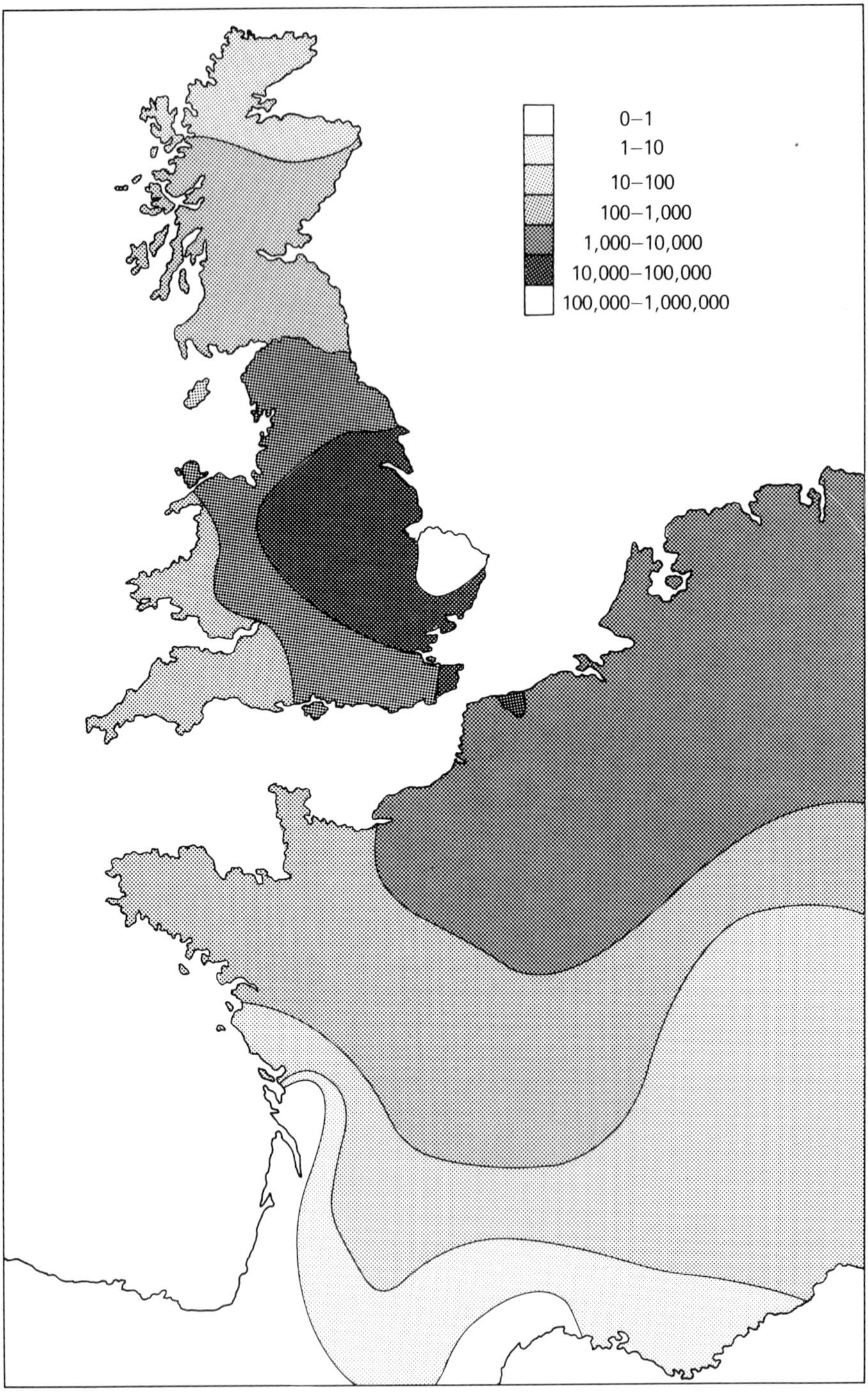
0–1
1–10
10–100
100–1,000
1,000–10,000
10,000–100,000
100,000–1,000,000

(a)

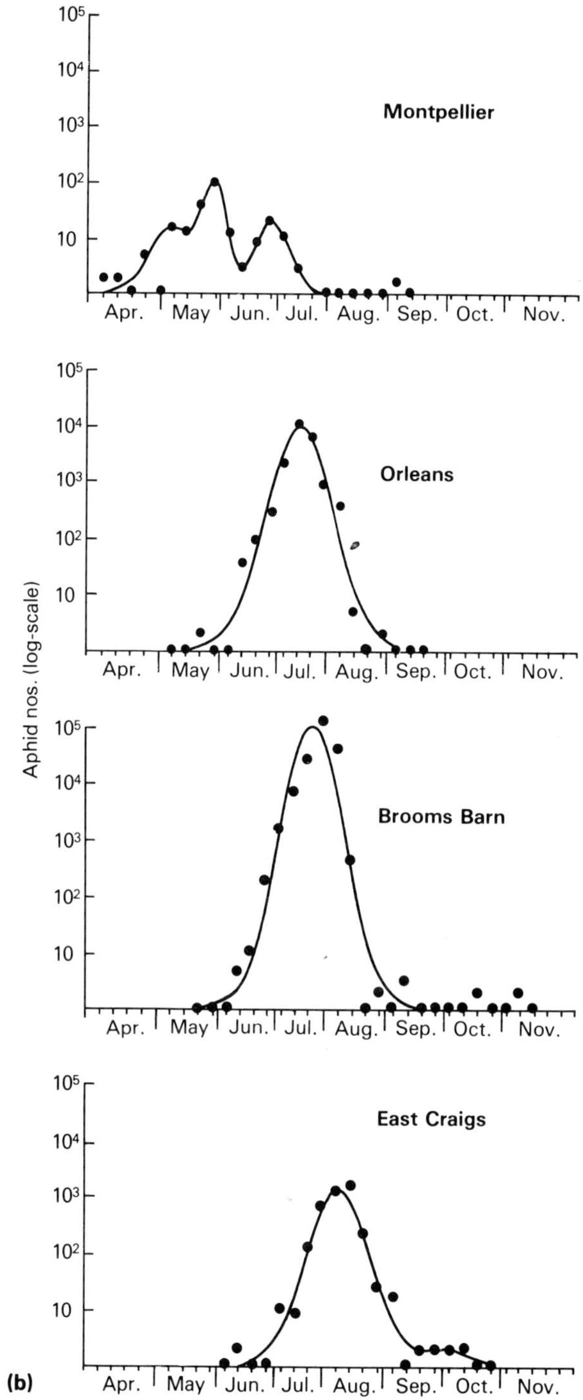

Fig. 2.8 (a) Trap catches of *Metopolophium dirhodum* in Europe from 26 July – 1 August 1979, mapped using SYMAP V programme
(b) Population curves of *M. dirhodum* caught in suction traps at Montpellier, Orleans, Broom Barn and East Craigs. (From Dewar, Woiwod and Choppin de Janvry 1980)

Fig. 2.9 12.2 m high Rothamsted suction trap

Smaller traps with battery-operated lamps are now available for sampling mosquitoes (Service 1970, 1976) and can be placed in crops remote from a mains electricity supply. Catches in traps within crops reflect the activity of pest species attracted by the crop (Hartstack *et al.* 1973; Hartstack 1979), whereas light traps are usually sited away from crops and are used to study migration of moths and other insects. Sorting the catch needs trained

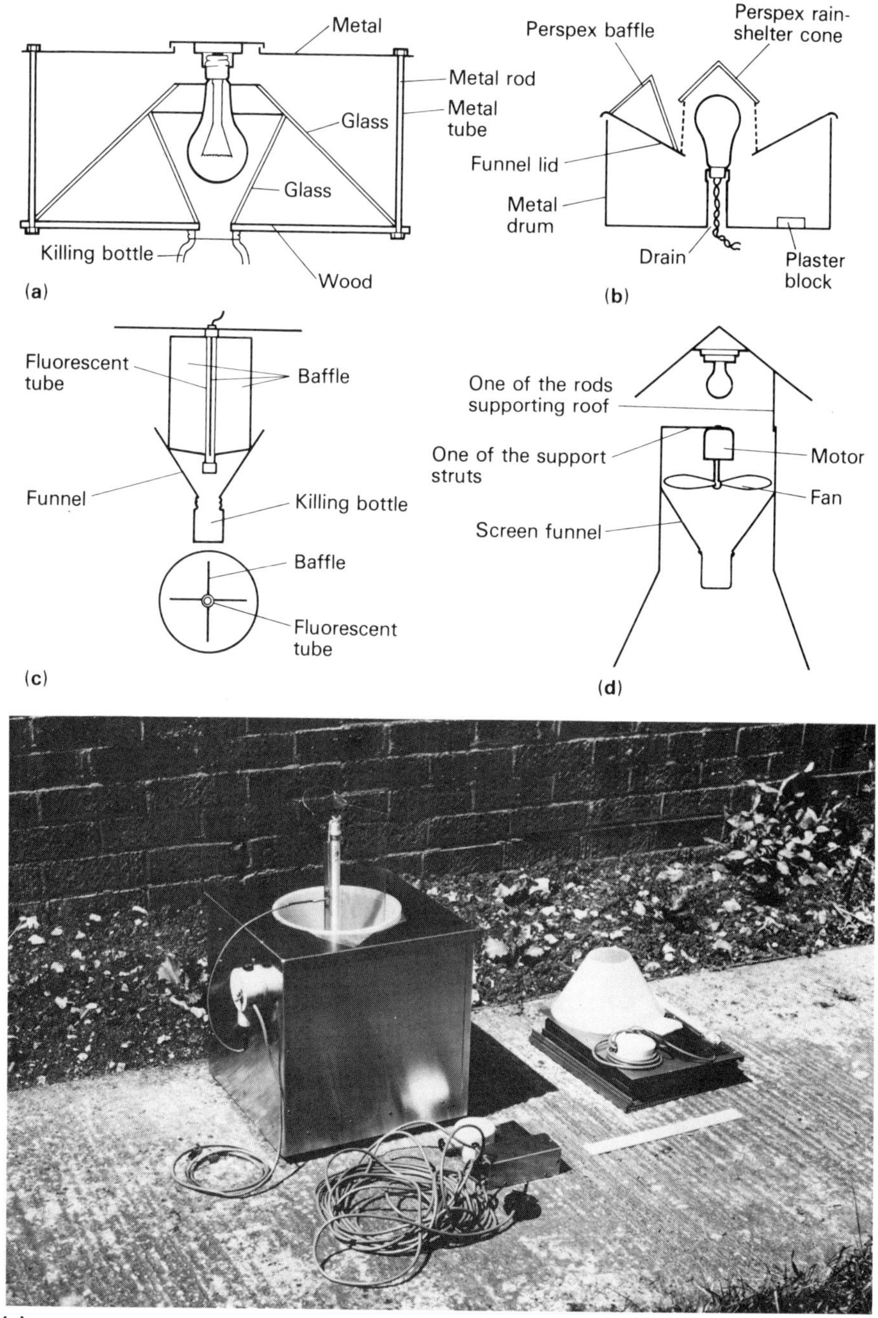

Fig. 2.10 Light traps
(a) Rothamsted (b) Robinson, (c) Pennsylvanian (d) New Jersey (From Southwood 1976) (e) Small 12 V Entech trap. (P.T. Walker photograph).

manpower so light traps are more useful for studying aerial density of insects where rapid identification is less important.

Efficiency of trapping is affected by moonlight (Bowden 1973a, b). Bowden and Church (1973) examined the interpretation of catches in relation to moon phases and showed that the effect of different light conditions on insect activity, abundance or behaviour influenced the size of catches. Also the proportion of a population which was trapped was affected by how far the trap was attractive. This variation in trap effectiveness with changing periods and amounts of moonlight obscures changes in activity and abundance, thus lower catches over the period of full moon often coincide with greater abundance or more activity (Bowden and Morris 1975). In Texas, Nemec (1971) reported a peak in the number of *Heliothis* spp. eggs on cotton just before a new moon.

Nevertheless, light traps have been used to improve timing of spray application, thus in East Africa a network of light traps has been used to monitor *Spodoptera exempta* migration to forecast where infestations are likely to occur (Brown, Betts and Rainey 1969). Even in East Africa, more selective trapping by using a pheromone is being introduced and hopefully will also identify areas in which *S. exempta* can survive for several generations or even throughout the year (Rose and Khasimuddin 1979). Pheromone traps were found to be better than light traps to detect pink bollworm in Africa (Marks 1976). Light trap catches have been improved when used close to a source of pheromone (Henneberry and Howland 1966). Light traps have also been used in conjunction with suction traps (Mulhern 1942) to catch weak flyers, and to survey migrations of the brown planthopper *Nilaparvata lugens* in Japan (Hirao 1979) but the relationship between catches and subsequent damage (hopperburn) is not clear. A catch of over 500 *N. lugens* in a light trap will result in a severe outbreak of hopperburn unless control is achieved. The threshold for hopperburn is thought to be between 50 and 100 hoppers (Table 2.1). A solar powered light trap has been used in the tropics.

Pheromone traps

Sex pheromones of lepidopteran pests are now used on an increasing scale to detect early infestations or to monitor populations and assist in the timing of pesticide application (Lewis 1977). The main advantage of pheromone traps over light traps is that normally only the target species is attracted and a trained entomologist is not needed to identify the catches. No power is required so pheromone traps are easier to operate and can be placed in remote areas.

Careful studies of the behavioural response of the insect in relation to the shape of the trap and method of pheromone release are needed to optimize the sensitivity of the trap. Most trap designs have been developed for their cheapness, convenience of handling, packing and checking or their durability in the field. In Tanzania 'Mushi' pheromone traps for *Spodoptera exempta*

Table 2.1. Brown planthopper immigrants up to 20 July, Chikugo, Japan

Year	Light trap catches	Hopperburn rating
1968	204	+
1969	2149	++
1970	177	++
1971	98	0
1972	31	+
1973	89	+
1974	97	++
1975	266	++
1976	191	+

Hopperburn rating: ++ overall hopperburn, + patch, 0 none.

are made from 4 gallon (18 litre) kerosene tins, containing water with detergent. The sides are cut open for moth entry (Rose and Khasimuddin 1979). Lewis and Macaulay (1976) and Wall, Greenway and Burt (1976) examined different traps for pea moth (*Cydia nigricana*) using 0.1 mg of one of its synthetic sex attractants, such as (E,E)-8,10-dodecanyl acetate in rubber dispensers. The simple triangular trap 180 mm long with sides of 20 gauge aluminium sheet and a removable base of 'Makrolon' (polycarbonate) coated with sticky material caught ten times more moths than the least successful designs. Larger catches were probably due to the well-defined plume of odour emitted from triangular traps aligned with the prevailing wind. Triangular waxed cardboard traps such as the Sectar (3M) fold-up adhesive trap have been used for a number of other insects including the pink bollworm *Pectinophora gossypiella* (Foster, Staten and Miller 1977).

Sensitivity of these traps is affected by wind conditions, the height of traps in relation to the crop and the adhesive used. For monitoring pea moth, traps were set at three-quarters crop height. Trap catches were seven times greater when 'Bird Tanglefoot' (The Tanglefoot Co., Grand Rapids, Michigan, USA) was used compared with 'Stickem Special' (Michel and Pelton Co., California) (Lewis and Macaulay 1976).

Unpainted galvanized metal funnel traps each with a perforated container have also been extensively used to keep moths alive for marking and release (Marks 1978), and yellow plastic funnel-traps with a 2 cm gap between the funnel and its lid were shown to be as effective as water traps for trapping *Spodoptera littoralis* (McVeigh, Paton and Hall 1979).

Hartstack, Witz and Buck (1979) described an inverted cone trap for *Heliothis virescens*, but in India, *H. armigera* has been very effectively trapped in a simpler trap consisting of a circular top over a funnel to which a polythene bag is attached to collect the moths; catches were improved with a smaller funnel attached to the lower surface of the top (Reed 1981).

In trials using 5 μl doses of gossyplure, pink bollworm moth catches were more than double those when using enclosed omnidirectional or large 60 cm diameter flat plywood disc survey traps (Fig. 2.11), both supplied with

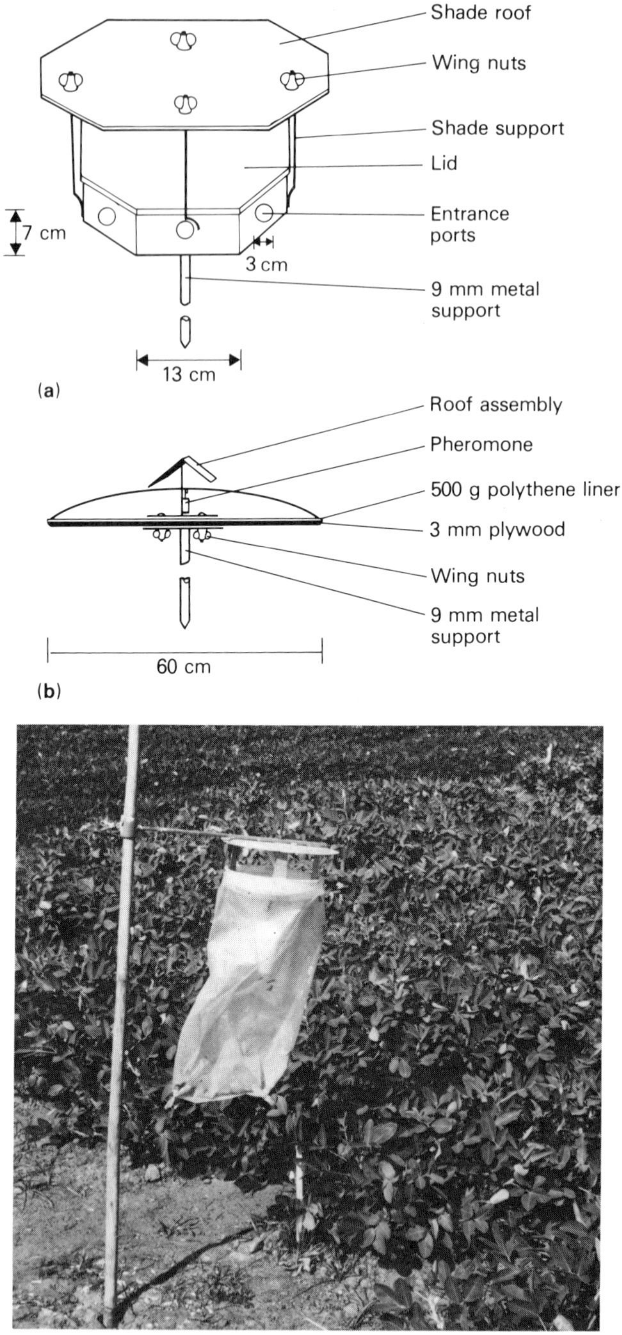

(c)

Fig. 2.11 (a) Omnidirectional pink bollworm pheromone trap. (From Marks 1976b)
(b) Flat plywood survey disc
(c) *Heliothis* pheromone trap in India

Table 2.2 Effect of trap height on catch of pink and red bollworm moths in trials in Malawi. (From Marks 1976b)

Trap height (in)	Pink bollworm moth catch	Red bollworm moth catch
0.5	26 a	26.33 a
1.0	125 b	16.03 b
1.5	31 c	2.83 cd
2.9	26 cd	1.47 de
2.5	19 cd	—
3.0	10 d	0.83 e

Treatments followed by the same letter are not significantly different $P = 0.05$.

Tanglefoot adhesive compared with Sectar traps (Marks 1976b). The ideal vapour dispenser has a uniform release rate preferably limited to periods of biological effectiveness, but none of the existing dispensers is satisfactory since emission rate rises with temperature (Nightingale 1979). Catches of red and pink bollworm moths were greatest when traps were not more than 0.5–1 m above the ground (Table 2.2). Metal water traps (60 × 60 × 7.5 cm deep) (Campion *et al.* 1974; Marks 1976a and b, 1978) have been used extensively in preliminary evaluation of pheromones, and caught more red bollworm moths than other trap designs (Marks 1978), but are heavy and expensive (Table 2.3).

Table 2.3 Catches of red bollworm *Diparopsis castanea* with traps baited with 1 mg Dicastalure. (Marks 1978)

Water trap with vane	24.2a
Water trap	21.9a
Plywood disc trap with sticky liner	14.2b
Metal funnel trap	6.1c
Double disc trap	5.5c
Sectar trap	0.2d
Tubular trap with sticky liner	0
Metal funnel trap covered by cylinder	0
Sticky string trap	0
Inverted cone trap	0

In addition to the design and position of the trap, pheromone trap efficiency is very dependent on having the appropriate mixture of the various chemical components or isomers and choosing a suitable vapour-dispensing mechanism.

Polyethylene vials (36 × 16 mm with 1.5 mm thick walls) and smaller Beem vials (19 × 9 mm) have been widely used and with the pheromone gossyplure catches were greater than when narrow gauge teflon capillary tube was used (Marks 1976b). When trapping *Spodoptera* polyethylene vials were better than rubber septa or a PVC laminate, but conversely catches of *Heliothis armigera* were greater when a rubber dispenser was used (Kehat *et al.* 1980) and in those trials, the addition of a third component of the pheromone improved the catch (Nesbitt *et al.* 1980).

Differences in the number and proportion of minor components may occur in geographically separated races of the same species (Campion and Nesbitt 1981). In *Spodoptera littoralis*, marked differences in the pheromone blends have occurred in insects originating from Crete, Egypt and Israel (Campion *et al.* 1980). Forty to forty-five per cent of the (2, E) isomer was the correct proportion mixed with the (Z, 2) isomer of gossyplure to give the best catches of pink bollworm (Marks 1976b).

One of the problems with pheromone traps is relating the numbers of moths trapped with subsequent damage to crops. Trap catches are relatively good at low populations, but the proportion trapped is expected to decrease when larger populations are present in the vicinity of the trap. Marks' (1977) attempts to correlate red bollworm moth catches with scouting data, were unsuccessful as oviposition was affected by crop growth, some fields being more heavily infested than others. *Spodoptera littoralis* trap catches were greater in lucerne and vegetable fields attractive to the pest than in other crops (McVeigh, Paton and Hall 1979).

Timing of sprays with pheromone traps has been successful when sprays against pea moth larvae are timed about nine days after ten male moths are detected in either of two traps per field (Lewis and Sturgeon 1978; Macaulay 1977).

In apple orchards, thresholds below which treatment is unnecessary are five codling moths *Cydia pomonella* per trap per week and twenty fruit tree tortrix *Archips podana* moths per trap per week (Alford *et al.* 1979). The possibilities of using pheromones to control pests by mass trapping and by disruption of mating behaviour are considered in the next chapter.

Other attractant traps

Various types of trap have been devised in which an attractant is used to lure specific pests or those attracted to a particular crop. Seven of the insect pests of brassicas may be attracted to traps baited with allylisothio cyanate (Finch 1977) and can be detected at lower densities than is achieved by crop inspection. In olive groves *Dacus oleae* is attracted to McPhail traps containing either protein hydrolysate or ammonium sulphate solution. Borax can be added to preserve the attractant over a longer period. Insecticide sprays have been recommended if more than seven olive flies are collected per trap per week. These traps collect both sexes in contrast to pheromone traps.

Soil samples

Examination of soil cores is often useful in assessing overwintering populations. Southwood (1966) describes soil corers and methods of sieving flotation and soil washing to extract the organisms. The most widely used methods were described by Salt and Hollick (1944) (Fig. 2.12), to assess wireworm population before sowing. The process is laborious but the data can be helpful to the farmer in deciding whether to plant a particular crop and whether control measures are needed. Hughes (1960) estimated

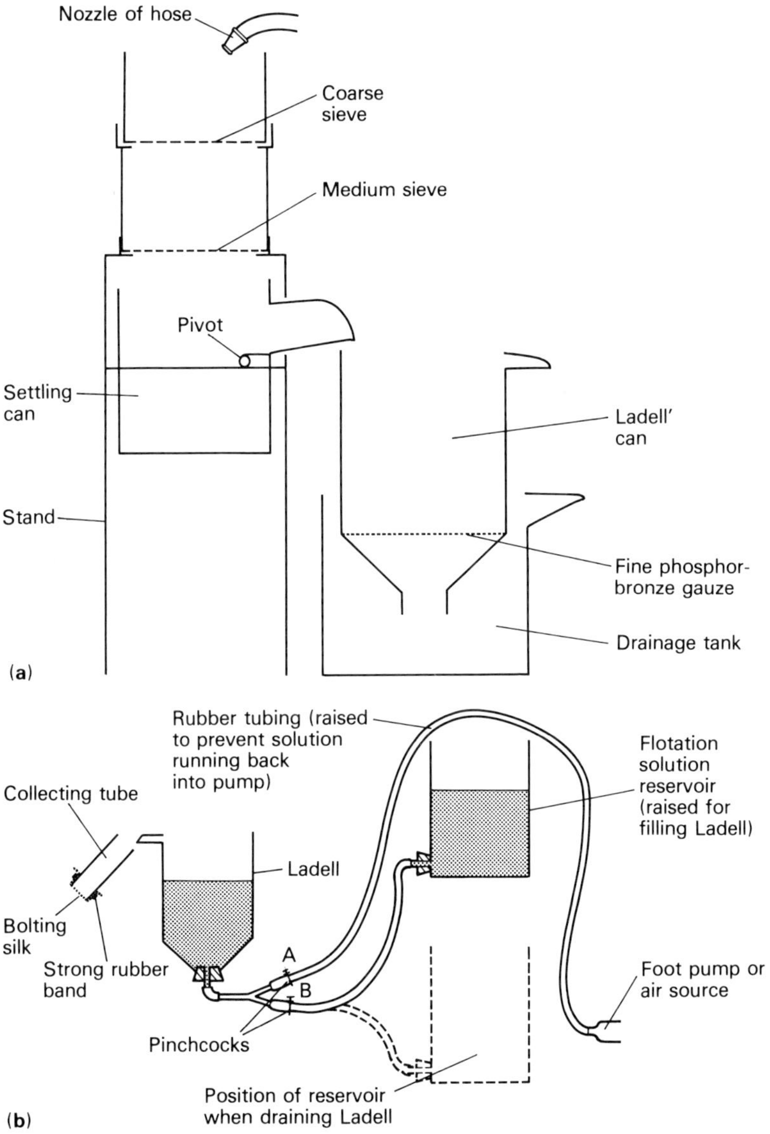

Fig. 2.12 (a) Soil-washing apparatus. (Modified from Salt and Hollick 1944)
(b) Ladell can and associated equipment during air-agitation phase of flotation. (From Southwood 1966)

numbers of cabbage root fly pupae in this way. Tunstall (1968) examined large samples of soil for bollworm pupae by digging up 0.9 × 0.9 m samples to a depth of 15 cm and after dry-sieving, the soil was washed through a 0.5 cm wire screen.

Phenological monitoring

In contrast to sampling for insects, the timing of fungicide applications is often related to the stage of plant growth and meteorological conditions. 'Beaumont' or 'Smith' periods are when humidity and temperature are favourable for blight disease development on potatoes (Beaumont 1947). Thus blight can be expected approximately 10 days after a period of 2 days if the temperature is over 10 °C and the relative humidity is greater than 75 per cent. Similarly 'Mills' periods can be used for apple scab (Table 2.4). Leaf wetness also influences development of coffee berry disease and coffee leaf rusts. However, Polley and Clarkson (1978) point out that meteorological data alone are not suitable for predicting optimum timing of sprays against cereal diseases owing to other factors affecting host–pathogen interactions. Sticky cylinders (Jenkyn 1974) (Fig. 2.13a) or the Hirst trap (Hirst 1952) can be used to collect spores for epidemiological studies. Bartlett and Bainbridge (1978) describe other methods for sampling micro-organisms in the atmosphere.

Table 2.4 Mills periods – examples of duration of leaf wetness at different temperatures

Mean temperature (°C) over period	5.6	7.2	10	11.7	15
Hours of leaf wetness	30.0	20.0	14	11.0	10

In the control of sigatoka disease on bananas in the Caribbean, a Piche evaporimeter has been used to measure the relative dryness of the prevailing weather. As the interval between sprays can be increased in dry weather, timing sprays on evaporation data is effective when field sanitation is good and there is no delay in fungicide application (Cronshaw 1982).

Rain-dispersed spores can be collected under small bushes and trees using a gutter draining into a container (Waller 1972) (Fig. 2.13b).

Assessing weed populations

Weed control is essential and farmers will cultivate their fields or apply herbicides irrespective of weed populations. Nevertheless, farmers or their advisers can walk across fields and assess weed populations visually before deciding on future cultivations or herbicide treatments. The history of individual fields is an important factor, so counts of weeds in random quadrats

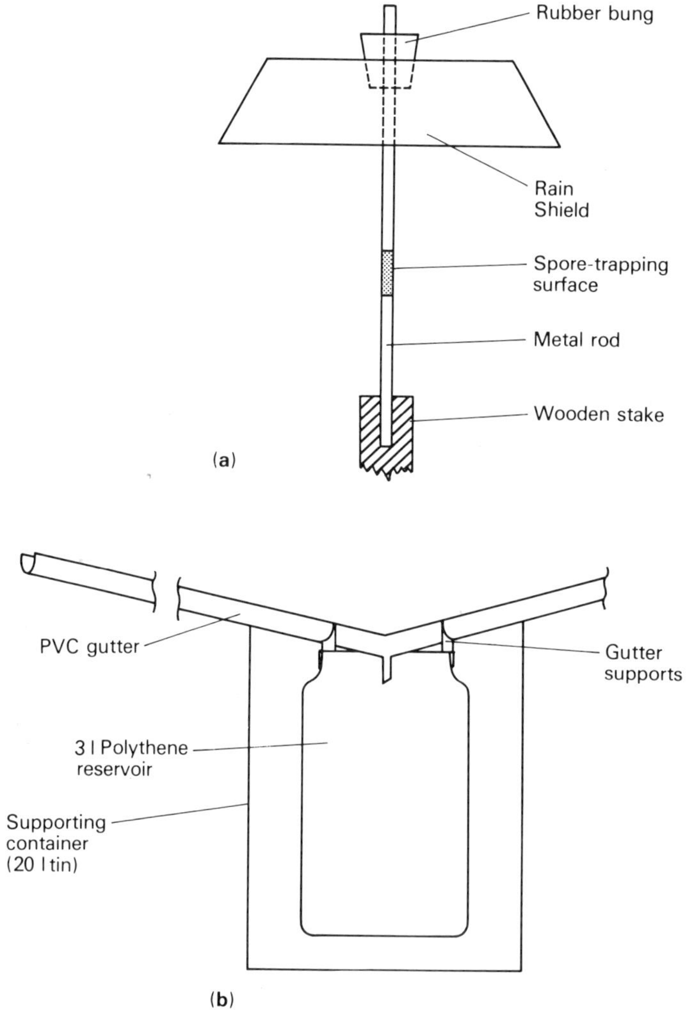

Fig. 2.13 Simple spore traps
(a) Wind-blown spores. (From Jenkyn 1974)
(b) Rain-dispersed spores from small bushes on trees. (From Waller 1972)

provide data for comparisons between seasons. Such counts are very important in trials work. Moss (1981) and Wilson (1981) give techniques for the assessment of black grass *Alopecurus myosuroides* and wild oat *Avena fatua* respectively. In addition to simple counts of weeds early in the season to assess weed competition, later counts are important to indicate weed seed production and estimate changes in populations. Assessing seed production itself is more time-consuming and of limited value unless weed viability is also measured. Weed population assessment is likely to play a greater role in weed management programmes in the future (Fryer 1981).

Monitoring vs Forecasting

A distinction should be made between forecasting the possible level of pest infestation several months in advance and short-term monitoring of pest levels or crops, after which action is usually taken within the week. Way and Cammell (1977) give economic thresholds for *Aphis fabae*, a species which can be readily sampled to give accurate predictions of crop infestations in the UK up to 5 months before they happen (Table 2.5). Such long-term forecasting is unique, since many factors usually influence the development of pest populations. Even decisions based on short-term monitoring can be upset by changes in weather, e.g. rain may wash-off eggs or wind and wet conditions reduce oviposition. Measurements of temperature in terms of cumulative day degrees or the sum of the average number of degrees above the threshold per day can be used in some cases to predict the onset of an infestation regardless of variable weather.

In this chapter a number of sampling techniques has been described, but more information is needed to relate pest numbers with crop damage under different weather conditions. Only by building-up local experience will it be possible to devise appropriate criteria to time control treatments. Nevertheless, it should be possible to reduce the number of pesticide treatments as such information is accumulated, and threshold levels adjusted.

Some examples of monitoring procedures are given in Table 2.6.

Table 2.5 Economic injury thresholds for black bean aphid (*Aphis fabae*) attacking spring-sown field beans in the UK. (From Way and Cammell 1977)

Time of monitoring	Monitoring technique	Economic injury threshold	Interval before insecticide needs to be applied
Eggs	Twig samples from spindle tree	1.1 eggs/100 buds	Above 5 months
Migrant alatae during dispersal	'Rothamsted' suction traps	5 alatae per trap before mid-June	1–2 weeks
Initial crop colonists	Crop sampling	5% of stems infested	Usually > 1 week

Table 2.6 Some examples of monitoring procedures and economic injury criteria. (After Way and Cammell 1979)

Crop pest	Field sampling:
Cereal aphids *Sitobion avenae* and *Metopolophium dirhodum*	5 aphids/head at beginning of flowering and numbers increasing (George 1975; George and Gair 1979)
Wheat bulbfly, *Delia coarctata*	Soil sampling in autumn 2.5 million eggs/ha but advice dependent on local conditions, sowing date and crop vigour

3

Which method of control?

Pesticides have been considered by many to be a panacea for the control of pests, but resurgence of pest populations, development of resistance to pesticides and concern about residues in the environment have led to a greater appreciation of the need to integrate different control techniques into a pest management system. The everyday natural regulation of populations means that very few organisms acquire pest status, so in the development of a control strategy, emphasis should be given to non-chemical controls whenever possible.

Reacting against the misuse or overuse of pesticides, many people argue that chemicals should not be used at all. In practice, application of a pesticide is often the only means of reducing a pest population rapidly, so improved methods of using pesticides are needed, to enable them to be integrated more effectively with other control methods.

Many alternative methods of control (Table 3.1) are used singly or in combination. The importance of these methods varies considerably in different pest situations, and only brief descriptions of some of the alternatives to pesticides are discussed here.

What are the alternatives to pesticides?

Plant resistance

Over the centuries, man has subconsciously selected seed from plants which survived pest attack, but hybridization and selection of cultivars which are resistant to particular pests or diseases have led to a greater understanding of why some plants are attacked more than others.

Resistance is defined as any inherited characteristic of a host plant which lessens the effects of parasitism (Russell 1978). Van der Plank (1963)

Table 3.1 Methods of controlling insect pests

1. Cultural

(a) Cultivation technique (ploughing, rotocultivation)
(b) Adjustment of planting date
(c) Adjustment of crop density or crop pattern
(d) Choice of cultivar (resistant varieties)
(e) Adjustment of irrigation and of fertilizer applications
(f) Use of barrier crops
(g) General crop hygiene
(h) Manipulation of harvesting procedures
(i) Manipulation or destruction of alternative hosts

2. Biological control

(a) Classical control – introduction of parasite or predator
(b) Inoculative – repeat introduction or augmentation of natural enemies
(c) Inundative repeated inoculations
(d) Preservation of natural enemies
(e) Use of pathogens (bacteria, fungi, viruses)

3. Interference

(a) Use of behaviour modifying chemicals (pheromones, kairomones)
(b) Use of insect hormones (juvenile hormones – moulting hormones)
(c) Use of antifeedants, repellants, or attractants
(d) Release of sterile insects
(e) Chemosterilants
(f) Genetic manipulation

4. Chemical control

(a) Use of selective insecticides
(b) Selective use of insecticides – timing in relation to pest and natural enemy populations, plant growth and/or meteorological factors, selective application technique, dosage rate

5. Integrated control

Deliberate integration of different methods to complement each other and minimize harmful effects in the environment

6. Pest management

Integrated control of pests on a regional scale requiring collaboration between farmers

suggested that there were two main types of resistance which can be distinguished in epidemiological terms. *Vertical resistance* occurs when a crop is very resistant to only certain genetic variants of a particular parasite, whereas *horizontal resistance* is generally polygenic and effective against a greater number of parasites, if not all to the same extent. Horizontal resistance is preferable as the variety is likely to remain resistant for a longer period, and is therefore more appropriately referred to as *durable resistance* (Johnson and Law 1975). Plant breeders have relied on vertical resistance, especially where the life cycle of the host crop is short and new cultures with differing resistance can be introduced rapidly. Horizontal resistance is more complex and requires selection from relatively large outbreeding populations.

One of several new varieties of rice is IR-36 which has multiple disease and insect from different genetic sources, which has been developed in South-East Asia.

Plant resistance or varietal control is now regarded as the cornerstone of insect pest and disease control, as it is compatible with other control techniques. Even low levels of resistance are important by reducing the need for other methods of control. The commercial development of a new variety can take years, and is expensive, so farmers have to pay a higher price for improved cultivars. The need for other control methods may be eliminated, or greatly reduced and no extra labour costs are involved, so the overall cost of plant resistance may be less than other methods. Even subsistence farmers can achieve better yields when improved seed is distributed nationally by government agencies, so an important part of the work of international institutes of agricultural research is concerned with breeding programmes, and consequently the maintenance of germplasm collections of major crops (Harlan and Starks 1980). Seed of varieties which are largely susceptible, may have some genetic resistance, so care must be taken to preserve important genetic material for future use.

Painter (1951) classified plant resistance into three types: non-preference, antibiosis and tolerance – but van Emden (1972) has differentiated nine mechanisms by which varietal control can operate (Fig. 3.1). More than one mechanism may operate in a resistant variety, and as a resistance mechanism for one pest can be an attractant or favourable for another, a compromise has to be made to satisfy agronomic requirements. When a number of pests and diseases occur, varietal control of at least one or two pests is essential, particularly to reduce the reliance on too many pesticides and to enhance biological control.

The use of different resistance mechanisms in cotton provides a good example of the use of varietal control as an essential part of a pest management system. Gland cells in the leaves of cotton produce gossypol which prevents many insects from attacking the crop. When plant breeders attempt to grow glandless cotton varieties, the plants are attacked by chafers and other insects normally found on maize and other crops (Gillham 1965). Increased levels of gossypol give greater resistance to many cotton insects, including *Heliothis* spp. (Lukefahr and Houghtaling 1969) though the presence of gossypol actually attracts the boll weevil *Anthonomus grandis*.

Cotton could be established in many parts of Africa by growing pubescent varieties (with hairy leaves) resistant to jassids *Empoasca* spp. (Parnell 1925; Parnell, King and Ruston 1949). The success of schemes involving jassid resistance has been followed by selection of varieties also resistant to bac-

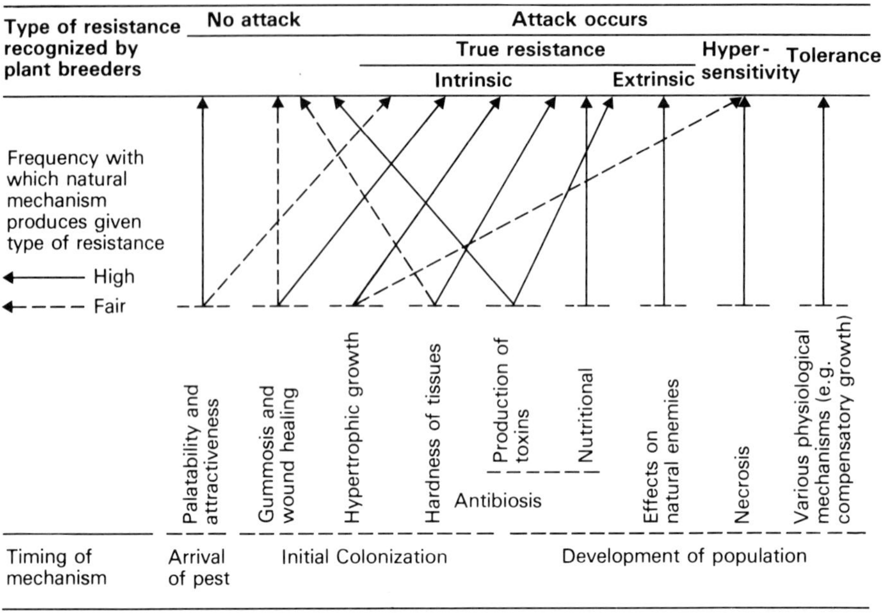

Fig. 3.1 A classification on plant resistance based on host plant influences on insect population. (From van Emden 1972)

terial blight (*Xanthomonas malvacearum*). Attempts to introduce larger boll, heavier-yielding glabrous (smooth leaved) varieties have tended to fail unless the period of insecticide treatment is extended throughout the season, so breeding trials need to be repeated without sprays to ensure that jassid resistance is maintained when selecting for higher yields or better lint quality. Nevertheless, in some countries glabrous varieties are preferred because hairy plants are more susceptible to whiteflies (*Bemisia* spp.) and because mechanical harvesting is easier.

Selection of a resistant variety on the basis of non-preference needs to be investigated on a large scale as differences in oviposition observed on small experimental plots, where there is choice, may not be repeated on a farm scale. Other morphological characteristics which are being studied include 'okra' leaf and 'frego' bracts (Schuster 1980). These varieties with more open foliage and more exposed buds provide less protection to pests from natural enemies and, if sprays are applied, allow greater penetration of droplets to increase insecticide deposits on bracts and bolls.

More determinate varieties which flower over a shorter period have been successfully developed in Texas, e.g. Tamcot SP–37 (Phillips, Gutierrez and Adkisson 1980; Walker 1980), so that even if insecticides have to be applied, the period over which protection is required can be shortened. Strict control of irrigation and fertilizer application to avoid prolonged bud production and lush growth with excess nitrogen is also essential to limit the period of growth exposed to pest attack. The ability of cotton plants to tolerate pest attack by continual growth as long as soil moisture and temperature are adequate was referred to in Chapter 1.

Despite several resistance mechanisms operating yield losses from cotton crops are still large, so varietal control must be used in conjunction with cultural, biological and chemical controls. Many other examples could be given for different crops, like rice where early-maturing varieties may escape late stem-borer attack.

In the same way that pests resistant to a pesticide are selected, so total reliance on varietal control is unacceptable. If selection pressure is considerable, a particular biotype which can survive is liable to be selected very rapidly. Chiang and French (1980) suggest that tolerance to a pest should be regarded as a short-term pest management tool. Corn rootworm *Diabrotica virgifera* caused greater damage to a susceptible maize variety, but more rootworms survived on a tolerant variety. The implication is that continued large-scale plantings of a tolerant variety would be more conducive to a pest population build-up than with susceptible varieties (Pathak 1970).

If only one variety is sown and resistance breaks down, rapid spread of a disease could be catastrophic, so Wolfe (1981) has suggested that several varieties with different resistance characteristics should be sown within one field to reduce the spread of a pathogen. Chin and Ajimilah (1982) have demonstrated that mixtures of rice varieties may be an effective method of controlling the blast disease 'Jaya'. A variety with good-quality grain but high susceptibility to blast can be grown by protecting it with Sekencang and MR25. Partial inhibition of hessian-fly larvae on resistant wheat is sufficient to reduce the pest to minor status in areas of the USA where the 'resistant' varieties have been used on a wide scale.

A similar classification of resistance to pathogens refers to disease escape, poor establishment of a pathogen, slow development of a pathogen, resistance to sporulation and crop tolerance.

Cultural control

Cultural controls manipulate cropping techniques. Many methods have been developed traditionally, not specifically for insect pest, pathogen or weed control and depend on local resources.

Crop rotation must be one of the oldest and most widespread practices to reduce pest problems. The separation of a pest from its host plant by growing an alternative crop remains an essential part of pest management, particularly for pests which are difficult, if not impossible, to control by other methods. This is particularly true of soil pests, since incorporation of a chemical into the soil is usually inadequate, and too expensive; thus nematodes are more effectively controlled by rotations.

Root knot nematode *Meloidogyne incognita* on tobacco is controlled in seedbeds by fumigation with methyl bromide, but seedlings are transplanted into areas which were previously grass leys for two to four years. Suitable grasses include *Chloris gayana*, *Panicum maximum* and *Eragrostis curvula*. With other crops such as bananas, a crop rotation is not possible, however.

Several rice crops may be grown in one field each year particularly under irrigation, so pest damage can increase in these areas unless biological control or some other factor limits pest population growth. Where alternative crops can be sown in the cooler or dry season, rice pests are reduced, especially if rice stubbles are ploughed soon after harvest. When there are several rice crops, synchronous planting is recommended rather than having prolonged sowing dates, as pests from early sown crops are liable to cause more serious damage to late sowings.

Cultivation of crops has been primarily for weed control; indeed, the earliest shifting agriculture involved opening up new land, when the farmer could no longer cope with the weed problem and soil fertility had deteriorated. Sometimes cultivation is necessary to break-up a hard pan, but it can have an adverse effect by breaking-up rhizomes and bringing-up new seeds while some cultivators increase compaction problems by merely scraping the surface.

With some crops the energy cost of ploughing and resowing has increased greatly, so some farmers have adopted minimum or zero tillage systems, relying upon use of herbicides to maintain adequate weed control. Direct drilling of cereals has expanded so far without bringing about any major changes in pest damage. Infestations of wheat-bulb fly have been reduced because the soil is never fallow, but winter cereals should not be sown too early. The incidence of some diseases such as 'take all' can increase initially, but the subsequent build-up of soil microflora exerts sufficient biological control of infection from ascospores. Care must obviously be taken to avoid applying chemicals which could interfere with such an equilibrium.

Minimum tillage systems can lead to an improvement in soil structure and reduction in loss of soil due to erosion, but changes in weed flora occur and require careful selection of herbicide treatments. After several years land sown to cereals may need to be ploughed and a break crop such as sugar beet or potatoes grown in rotation to control certain weeds and nematodes.

On many farms, particularly in the tropics, several crops may be grown in a single field. In some cases the crops are completely mixed, but more often they are segregated into individual rows or strips. Ecological diversity is important as the level of pest incidence on any one crop in the system is usually less than when that crop is grown in pure stand. Unfortunately, as traditionally practised, the system is labour intensive, especially for weed control and harvesting. Recently the use of mixtures of varieties with different levels of resistance (Wolfe 1981) have been suggested to deter the spread of a particular pathogen which would occur if one susceptible variety was sown throughout a field. The varieties chosen would need to mature at the same time and have sufficiently similar quality to allow modern combine harvesting. On small-scale farms, further development of strip-cropping is a possibility where placement of a pesticide selectively on an individual crop can be achieved without contamination of neighbouring strips (see p. 125). A danger when strips are grown is that some pests (e.g. grasshoppers) may become more serious due to an increased edge-effect. However, if one of the crops in the system acts as a trap crop, the overall damage may be reduced, especially if it is easier to control a pest population on the trap crop. In combinations of sorghum and cotton, application of a virus to control *Heliothis* may be more effective on sorghum as the virus remains active over a much longer period than when applied on cotton (Roome 1975; Roome and Daoust 1975).

Pest attack is often affected by time of sowing and spacing of crops; thus close-spaced (2.5 cm within rows 90 cm apart) early-sown groundnuts had fewer alate *Aphis craccivora* and less rosette virus disease and yields were better than when groundnuts were intersown with beans to reduce spread of the virus (Farrell 1976a, b). Similarly, early-sown tobacco is less at risk to aphid and virus infestation than late-sown tobacco (Legge 1966). Conversely, early-sown winter barley crops can be more exposed to aphid infestation in the autumn, especially in South-West England, and can be severely damaged by barley yellow dwarf virus.

In many areas where sowing is dictated by the onset of the rains, late-sown crops yield poorly, even in the absence of pest attack, so revenue is usually inadequate in relation to the costs for any pest management. In Nigeria, yields of cotton sown late after weeding has been completed on cereals and legume crops are generally too low to justify use of insecticides. Thus sowing at the optimum time agronomically is needed to promote potential yields worth protecting. Early sowing can be combined with a trap crop when irrigation is available. In South-East Asia a quarter of a rice crop can be planted 20 days early in the area surrounding the main crop, so it attracts more colonizing brown planthoppers than the latter during the period up to 75 days after transplanting. Treatment of the trap crop area can reduce the need to treat the whole area until later and allow greater survival of natural enemies in the main crop. Early-sown cotton has been used in parts of the USA to trap boll weevils in the same way. This technique could possibly be improved by applying the pheromone Grandlure to the trap crop. Melon and other cucurbits can be protected partially from melon flies by planting a trap area with maize treated with insecticide. In this way the susceptible crop of cucurbits does not have any pesticide residue and the few flies which attack fruit can be controlled by natural enemies.

Plant spacing often presents a dilemma for the grower. When rice is grown, a low tiller density favours stem-borers, but conversely a high tiller density or closer spacing encourages more planthoppers. Penetration of an insecticide spray is better with a low tiller density, but the potential for plant compensation is greater with more plants. A compromise is to plant rice on a 5 × 40 cm spacing, preferably with the rows orientated parallel to the path of the sun to reduce shading and the relative humidity. The wider spacing allows easier access for the farmer to inspect his crop and apply a pesticide if needed.

Another dilemma in relation to spacing is that farmers often use an excess of seed to get a crop established and allow for some loss of seedlings. Precision drilling reduces the amount of seed sown and the need to thin crops, but increases the risk of subsequent gaps due to pests and diseases.

End-of-season crop sanitation is often a vital method of controlling pests and diseases which could otherwise carry over to the next crop. Burning of cereal stubbles is important in continuous cereal production, since many pests including hessian fly and wheat-stem sawfly are destroyed. In most cotton-growing countries the date for uprooting or shredding and ploughing-in crop residues is proscribed by law in order to control pink bollworm *Pectinophora gossypiella* and bacterial blight *Xanthomonas malvacearum*. In India, goats were used to forage cotton fields after harvesting to remove bolls infested with pink bollworm. Care must be taken to make sure animals do not enter pesticide-treated fields until sufficient time has elapsed after the last treatment.

Removal of damaged fruit is also very important in many perennial crops, particularly if harvesting is over a period of several weeks. Phytosanitary pruning of cocoa is important to reduce the spread of diseases such as *Phytophthora* and *Crinipellis*, and false codling moth *Cryptophlebia* can be checked by removing damaged citrus fruit. Garbage removal is an essential part of fly, cockroach and rat control and removal of abandoned containers, such as tyres, tin cans and even cars that collect water can greatly reduce urban mosquito problems.

Biological control

Traditionally, biological control has involved a search in the original habitat of the pest for a suitable parasite* or predator which can be released in the pest-infested area. A parasite is usually parasitic only in the immature stages of its host. It develops within or upon a single host, which is ultimately destroyed as the parasite larva completes its development. The adult is in most cases free-living and feeds on nectar, honeydew or sometimes fluids from other insects. Predators are relatively larger than the hosts, which are devoured or sucked dry very quickly. A single predator may kill hundreds of its prey. Characteristics of a good natural enemy are that it searches for prey even when it is at low population density, and has a high fecundity so that it can respond rapidly to an increase in the pest population. Some authors have argued that they should be host-specific, but there is an advantage for those pests that can maintain themselves on an alternative host when the preferred host is not available. The natural enemy should be reared easily and able to survive throughout the climatic range of the pest and be effective at the onset of a pest infestation.

* or parasitoid i.e. alternatively parasitic and free living.

The classic example of biological control is the suppression of cottony cushion scale *Icerya purchasi* by the ladybird *Rodolia cardinalis* in California ever since 1888, when 129 ladybirds were sent from Australia, by the entomologist Albert Koebele. The ladybird has been distributed now to over 35 countries and has successfully prevented destruction of citrus. A parasitic fly *Cryptochaetum iceryae* also released at the same time as *Rodolia*, remains important, especially in coastal areas of California (de Bach, Rosen and Kennett 1971).

Unfortunately if certain insecticides are applied to control other pests on citrus trees, adverse effects on the natural enemies can result in serious outbreaks of scales (de Bach and Bartlett 1951). Until recently chemical control was still required in Peru, against the rufous scale *Selenaspidus articulatus*, but a search in East Africa has provided a chalcidoid *Asphytis roseni*. This was released from 1972–76 in unsprayed citrus orchards and has rapidly reduced the rufous scale populations to below the economic threshold (Beingolea 1977). The main advantage of the inoculative technique is the low cost when one introduction provides a self-regulatory system without further expense. Most successes have been confined either to geographical or 'ecological' islands and principally on perennial crops. Another classic example of inoculative biological control was the release of the tachinid fly *Ptchomyia remota* from Malaya, released in Fiji to control the coconut moth *Levuana iridescens*. Only 32,570 parasitized larvae were released and economic control was achieved within 2 years.

Success in many cases depends on searching for natural enemies in as many locations as possible to find one most adapted to its new environment (Caltagirone 1981). The Commonwealth Institute of Biological Control has provided expertise at a number of centres throughout the world to evaluate pest problems, select and screen potential species and subsequently breed sufficient for release. Examples of some recent successes are given in Table 3.2.

Inoculative release of an insect has also provided biological control of weeds. The first success followed the introduction of 2750 eggs of the moth

Table 3.2 Examples of successful introductions of biological control

Crop	**Pest**	**Country benefiting**	**Year of first introduction**
Potatoes	Potato tuber moth (*Phthorimaea*)	Zambia	1968
Citrus	Blackfly (*Aleurocanthus*)	Bahamas	1971
Sugar cane	White scale (*Anlacaspis*)	Tanzania	1971
Coconuts	Coconut hispid (*Promecotheca*)	Sri Lanka	1972
Vegetables pastures	Noctuids (*Mythimna*)	New Zealand	1973

Cactoblastis cactorum into Queensland from Argentina in 1925. In order to control the pricky pear cactus, *Opuntia stricta* was controlled very rapidly over millions of hectares and the success has been since repeated in many other areas. More recently, a weevil *Rhinocyllus conicus* has been released to control nodding thistle *Carduus nutans* in Canada (Harris and Zwolfer 1971) and in New Zealand (Ivens 1979). In Central Africa an aquatic grass-hopper *Paulinia acuminata* has become established on the water fern *Salvinia molesta* which had become a serious problem on Lake Kariba (Mitchell and Rose 1979).

Field surveys are needed to study the role of phytophagous organisms in the area close to the source of the weed species. A potential biocontrol agent may not look very promising when feeding on the weed in its native home, as numbers may be regulated by its own parasites and predators. In their absence the insect can be very effective, but before a candidate species can be released, its host specificity must be checked by means of screening tests (Wapshere 1975). If the potential biocontrol agent only feeds on the target weed (monophagous) or a few closely related species (oligophagous) it is considered safe to introduce without risk to crops and other non-target plants. Sometimes more than one agent is needed, thus an insect, mite and rust fungus have together been involved in the control of skeleton weed (*Chondrilla juncea*) in Australia.

Several factors can disrupt the balance between a pest and its natural enemies, thus preventing continued suppression of a pest. Cold or dry weather can interrupt biological control even in perennial crops, but the harvesting of annual crops is a major hindrance to biological control unless climatic conditions or irrigation allow continuous cropping or the natural enemy has a diapause or other method of remaining in phase with the pest. In the UK, the diamond-back moth, *Plutella maculipennis* is heavily parasitized by parasitic wasps *Horogenes* spp., which bridge the winter in another insect; thus *H. fenestralis* overwinters on *Swammerdamia lutarea* on hawthorn. Similarly, in California, the tiny mymarid *Anagrus epos* is important as a parasite of leafhoppers in vineyards but survives on an alternative host during the winter. This is a non-economic leafhopper *Dikrella cruentata* which occurs on blackberry *Rubus* spp. These bushes should not be removed from vineyards if farmers wish to promote biological control of leafhoppers.

An alternative method of biological control is to rear a predator such as *Chrysopa carnea*, or parasite, notably *Trichogramma* spp. for repeated releases in the field each year. Ables and Ridgway (1981) list entomophagous arthropods being used in trials for release on crops in the USA, western Europe, the USSR and China. Extensive use of parasites and predators has been reported from Russia (Beglyarov and Smetnik 1977) and China, but very large numbers need to be released; thus an estimated 45,000 *Trichogramma* per hectare would need to be released each 10-day period throughout an extensive cotton-growing area (Knipling 1979) to have a significant effect. Hagen, Sawall and Tassan (1971) have sprayed a food on vegetation to sustain the natural enemies and attract more into the crop, but comparatively little work has been done on inundative releases in the field. The application of a suitable attractant has helped to keep released parasites within a field; thus in some trials the pheromone of *Heliothis* on cotton helped to maintain *Trichogramma* populations in the same area. In some situations release of natural enemies over a critical period to augment the natural population may be what is needed to tip the balance in favour of biological control.

In glasshouses, spider mites are deliberately introduced on to every cucumber plant within a few days of planting via a small section of bean leaf bearing 10–20 female mites (Hussey and Bravenboer 1971). When the mean leaf damage index is 0.4 (Hussey and Parr 1963) two *Phytoseiulus persimilis* predatory mites are introduced on every second plant. By having the predatory mite with prey throughout the crop, adequate control of post-diapause mite populations which overwintered within cracks and crevices in the glasshouse can be achieved throughout the summer without further manipulation. *Encarsia formosa* is also released in glasshouses to control whiteflies *Trialeurodes*. Studies by Parr (1968) demonstrated the importance of an equilibrium between the pest and parasite throughout a crop to avoid an uncontrolled infestation on some plants, so both pest and, fourteen days later, the parasite are released between each pair of plants. Scopes, Randall and Biggerstaff (1975) describe a cage suitable for rearing phytophagous insects and their parasites (Fig. 3.2) but numerous techniques have to be developed to suit particular species. Spraying pesticides against other pests and diseases is liable to disrupt the interaction, so chemicals need to be screened for their effect on natural enemies (see p. 81). Some fungicides such as binapacryl are also acaricides and reduce the number of predaceous mites in orchards.

Many crops have more than one pest or disease, so further studies are needed to integrate several biological control systems. Lack of information on biological control usually necessitates one or more pesticides being applied, so the third approach to biological control, namely to conserve existing natural enemies is extremely important. This is achieved principally by avoiding pesticides for as long as possible particularly during the early part of a season reducing the number of applications and dosage, and applying them as selectively as possible. The choice of chemicals and application techniques are discussed more fully in later chapters, but in order to achieve selectivity, it is important to identify the significant predators in each area. Conservation of natural enemies, at field edges, e.g. in hedges is also needed.

Ants which tend scales, mealybugs and aphids also interfere with parasitism, so by placing a sticky band around trees to stop the ants, better biological control can be achieved.

The effectiveness of natural enemies can be demonstrated by various experimental techniques (de Bach and Huffaker 1971). These include use of mechanical barriers to exclude natural enemies; thus fine organdie cloth-sleeve cages can be used to stop parasite oviposition, but the presence of the

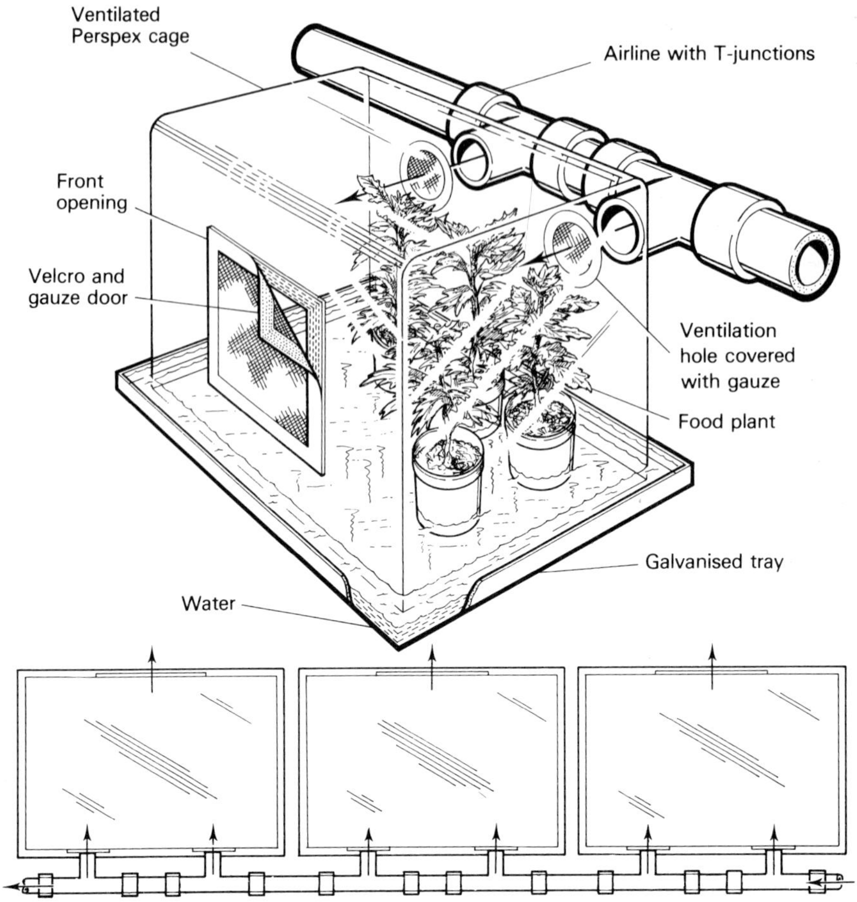

Fig. 3.2 Ventilated cage suitable for rearing phytophagous insects and their parasites on potted plants. (From Scopes, Randall and Biggerstaff 1975)

cages affects the microclimate for the pest.

Chemicals have been applied to remove natural enemies without any significant effect on the pest; thus DDT applied to citrus has less effect on the scales than its natural enemies, so a rapid increase in scale insects occurs (Fig. 3.3) (de Bach 1974).

Microbial pesticides

Pest populations are often regulated naturally by a disease, so another form of biological control is to use a pathogen such as a bacterium or virus against pests (Burges and Hussey 1971; Falcon 1976; Tinsley 1979, Payne, 1982). Pathogens can not only cause death, but can also lower resistance to attack by parasites, predators and other pathogens, and influence the devel-

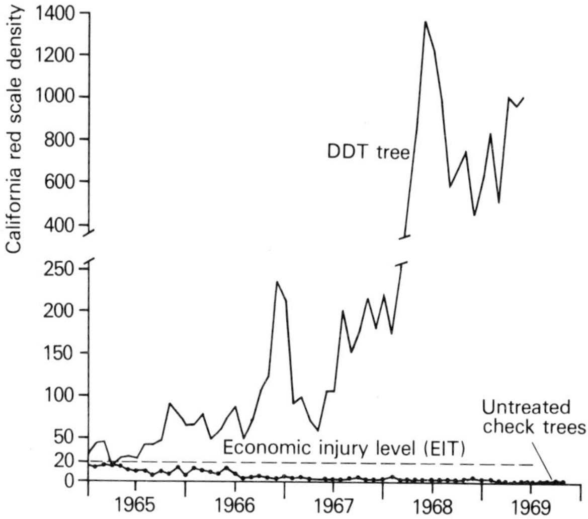

Fig. 3.3 Increase in numbers of red scale on citrus in California following application of a persistent insecticide reducing numbers of natural enemies. (From de Bach 1974)

opment, reproduction and susceptibility to chemical insecticides of an insect. The bacterium *Bacillus popillae* injected into the soil has successfully controlled the Japanese beetle *Popillae japonica* since 1942. The protein crystal of *Bacillus thuringiensis* is very toxic to certain lepidopterous larvae and a few diptera, but it is relatively harmless to other organisms. The spores and crystals have to be ingested, so this insecticide is most effective against leaf-eating larvae which have a strongly alkaline (pH 8.9) gut and an enzyme to dissolve the spore crystal and release the toxin. Control of stem-borers and fruit-feeders is very poor. Degradation of spray deposits and its slow action limit the use of *B. thuringiensis*, but new strains such as *B. t. isrealinsis* (serotype H–14) are very promising against mosquito larvae.

Some insect pests have been controlled by collecting diseased larvae and simply mixing their body contents in a homogenizer with water; the liquid is then filtered and sprayed on to crops. Nettle caterpillar *Darna trima* is controlled in oil palm plantations by spraying an aqueous suspension of triturated virus infected with *D. tirma* virus (Tiong 1982). This is no longer generally acceptable, owing to concern over the use of replicating microorganisms and the risk of spreading viruses similar to those infecting mammals and other organisms, such as the pox viruses. This has necessitated the use of purified baculoviruses, which only infect arthropods. The purified virus is subject to safety testing and the virus must be characterized so that it can be identified readily (Burges 1981; Harrap 1982). In Australia the specificity of *Baculovirus heliothis* had to be tested to avoid any risk of infecting *Cactoblastis cactorum* for fear of disruption control of prickly pear *Opuntia*. The nuclear polyhedrosis virus of the European spruce sawfly

Gilpinia hercyniae was introduced inadvertently into Canada and provides self-perpetuating control to maintain infestations below the threshold level for economic damage (Clausen 1978). Control of sawflies by baculovirus has been reviewed by Cunningham and Entwistle (1981). Persistence of a virus does not happen in all cases so a periodic application is usually required either once early in the season, e.g. to control Brassica pests after a soil application in Canada (Jaques 1972), or more routine applications as with chemical insecticides.

The main disadvantages of using microbial insecticides is that there is delay between time of application and mortality of the pests, and damage occurring in the intervening period may not be acceptable, although infested larvae often do stop feeding. In contrast to parasites and predators, microbials do not disperse, so good coverage of foliage is essential to ensure larvae come into contact with the pathogen. Movement of infected insects can be used to spread the baculovirus, e.g. with the coconut beetle *Oryctes rhinoceros* (Bedford 1980, 1981). This pest has been controlled successfully in many areas (Fig. 3.4) by releasing live beetles which had been immersed in a suspension of two ground cadavers/litre of water for 2–3 minutes and then allowing them to crawl through sterilized sawdust mixed with half a virus grub in 500 cm^3 of water. Microbial pesticides have to be protected against the effects of ultra-violet light (Gudauskas and Canerday 1968). As they are particulate the dilution ratio is limited if each droplet is to contain the pathogen. Feeding stimulants including molasses, and various plant extracts,

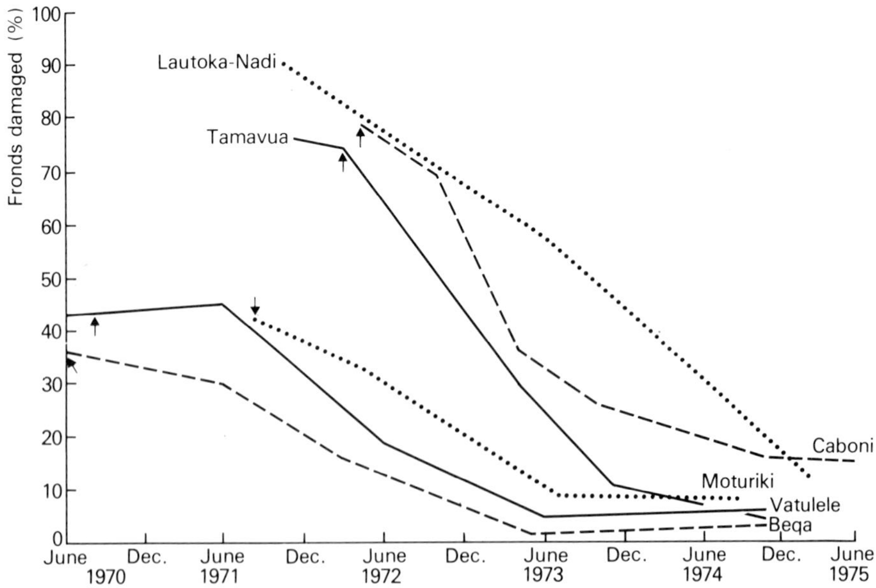

Fig. 3.4 Effect of baculovirus on palm damage by *Oryctes rhinoceros* at different localities in the Fiji islands. Arrows indicate date of virus introduction. (From Bedford 1980)

have been added to increase ingestion of microbials (see p. 131). Yield increases with baculovirus treatments have often been disappointing as alternatives to chemical insecticides on annual crops when in small field trials. Application of baculoviruses over larger areas as a component of a pest management system seems more appropriate in order to use their specificity and ability to enhance the effects of other control techniques. In Arkansas, increased parasitism of *Heliothis* spp. was reported following community-wide spraying of *Baculovirus heliothis*. Lower dosages are needed to control small larvae, but if larger larvae are infected, more virus will be produced and disseminated naturally.

Recently a fungus *Verticillium lecani* has been investigated as an aphicide (Hall 1981). It has proved very effective on chrysanthemums grown in glasshouses where the humidity can be raised by surrounding the plants with polythene sheeting to control day-length. In the tropics, aphid infestations are often controlled naturally by entomophagous fungi during the rains.

Beauveria bassiana is used in China mainly against European corn-borer *Ostrinia nubilalis* (Soper and Ward 1981), where it has been uniquely applied in forest areas using a mortar bomb (Way, personal communication).

Control of weeds with pathogens (mycoherbicides) is now being tried, especially in the USA.

Interference methods

Several methods of control rely on interfering with insects' behaviour. Lewis *et al.* (1980) has discussed the status and potential use of behavioural chemicals or semiochemicals in pest management. At present, marking pheromones and kairomones are being assessed as oviposition disruptants on crops, while kairomones may assist with the establishment of beneficial organisms, but interference with mating with pheromones is considered here together with interference with reproduction by release of sterile insects. The use of antifeedants is also discussed, but chemicals that interfere with the metamorphosis of insects are included in the section on choice of insecticides (see p. 76).

Pheromones

Pheromones are being investigated for control of pests in two ways; by mass trapping and communication disruption.

In areas where a moth population is low, and labour is available to position and subsequently maintain sufficient traps, removal of male moths by mass trapping may delay an increase in pest populations and provide a more economic method of control in comparison with the application of insecticides. The traps must be cheap, robust and simple to operate. So far, mass trapping has met with limited success (Campion 1983).

Following feasibility investigations into the control of *Spodoptera littoralis* in Egypt with 5 traps per hectare over 500-ha studies (Symmons and Rosenberg 1978), a reduction in egg masses of 77 per cent was achieved in 1980 (Campion and Nesbitt 1981) but not in 1981. Similar studies have been reported in Israel (Teich *et al.* 1979) and are in progress in Japan and India. Other studies are primarily concerned with trapping beetles in forest situations, e.g. *Ips typographus* is being trapped in Scandinavia and *Scolytus multistriatus* in the USA (Baker 1979).

The alternative control technique is to permeate the atmosphere with a pheromone to disrupt chemical communication between the sexes and reduce the frequency of mating. This disruption can be with the pheromone or with structurally similar chemicals which inhibit mating (Beevor and Campion 1979). Pheromones have been released from laminated three-ply dispensers (Beroza *et al.* 1974) and from hollow fibres, thus a gossyplure formulation inside hollow 'Conrel' fibres has been applied aerially on cotton in large-scale trials. Special apparatus is needed to dispense a slurry of chopped fibres mixed with polybutene polymers (Funkhouser and Las 1981). The aim is to provide early suppression of *Pectinophora gossypiella* so that insecticide treatments can be delayed. Brooks, Doane and Haworth (1979) reported a 67 per cent reduction in insecticide use with a higher yield. Timing of application is crucial. Other disruption trials have been with *Spodoptera littoralis* (Campion *et al.* 1976; Campion *et al.* 1980; Campion 1983), and in Malawi *Diparopsis castanea* oviposition could be reduced (Marks 1976a; Marks *et al.* 1978 (Fig. 3.5). These trials have shown the importance of uniform release of the pheromone over several weeks. This is achieved with a micro-encapsulated formulation which has been developed to protect the pheromone under field conditions from degradation in sunlight (Campion *et al.* 1980). The micro-encapsulated formulation is easier to apply than hollow fibres, as conventional spraying equipment can be used. The use of electrostatically charged sprays should increase the retention of the pheromone on foliage and reduce losses to the soil. Progress with pheromones

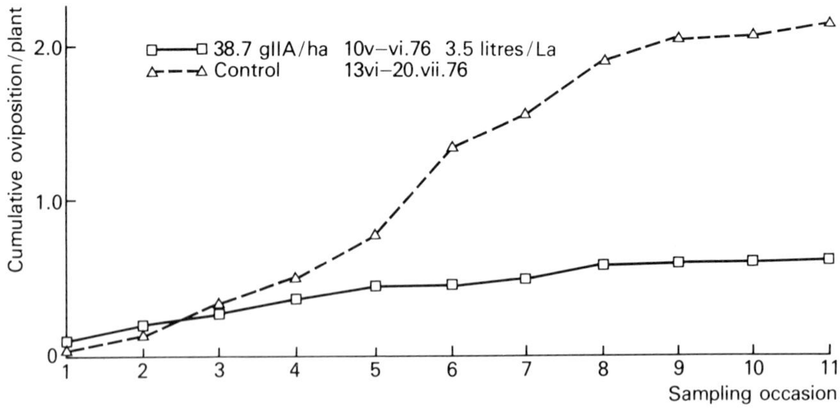

Fig. 3.5 Reduction in oviposition by *Diparopsis* on cotton in area treated with micro-encapsulated pheromone (IIA)

and similar chemicals will depend on a better understanding of the complex patterns of pest behaviour and the interaction of several chemicals which individually or in combination may affect different stages of a sequence of behavioural activities.

Sterile male release

Pest eradication is not a feature of a pest management programme, but several attempts have been made with autocidal control tactics to reduce particular pests over a large area.

The most successful use of release of sterile insects to compete with the wild population and reduce successful mating has been with the screwworm *Cochliomyia hominovorax*, an obligatory parasite of livestock in the southern USA, Central and South America (Richardson 1978). Pupae were irradiated at 7.5 krad. Initial success was when 400 sterile males per square mile were released each week on the isolated island of Curaçao. This was followed by releases in Florida through to Mexico with considerable success until setbacks occurred (Cueller and Brinklow 1973) possibly due to the sterilized flies being less competitive, or poor survival of the laboratory flies in the field. The sterile release technique has also been successful on the island of Rota in the Marianas, where the melonfly and oriental fruitfly were controlled.

Sterile male release studies have continued with several other major pests, including the Mexican fruitfly along the California–Mexico border and tsetse flies *Glossina* spp. in Africa, but so far no major control programme has been established. Major disadvantages to the method are the need to breed large numbers of insects for sterilizing; the technique is most effective on pest species where females mate only once, the population is low and is in some way isolated. Insecticides may have to be used to reduce a population before sterile insects can be released. Unfortunately, numbers of insects have to be bred unless a system can be devised for trapping sufficient numbers for sterilization.

Integrated control

A single control technique may be adequate in a pest management programme, but where several pests occur, more than one control tactic is often needed. Careful integration of several control techniques is required so that they do not oppose each other but harmonize to provide optimal cost-effective control over a long period rather than only a short-term solution (Fig. 3.6). As chemical control remains an important tool in pest management programmes, the selection of chemicals and their application is considered more fully in the following chapters.

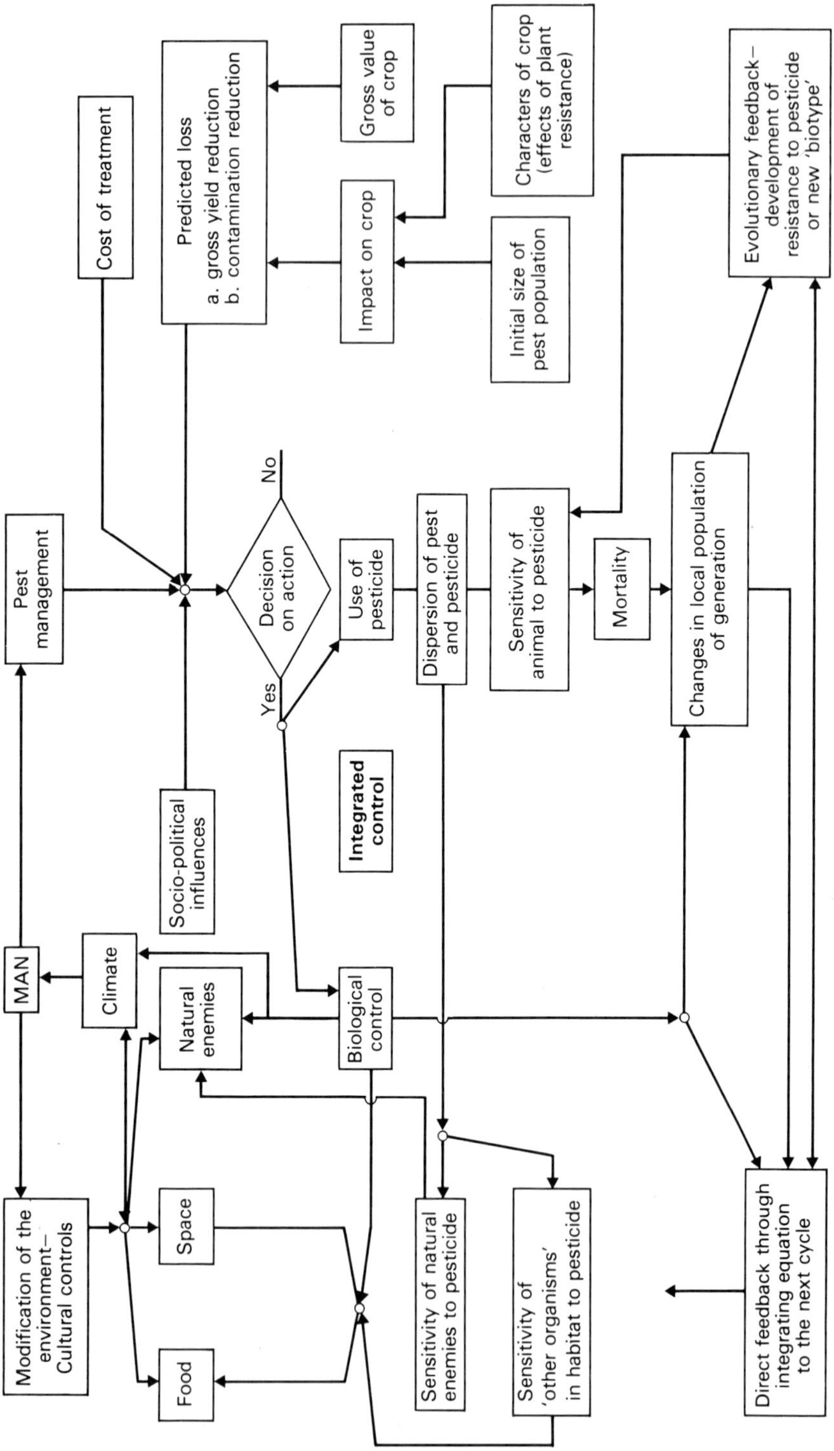

Fig. 3.6 General outline of factors in a pest management programme. (Adapted from Southwood 1968)

4

Which chemical?

The ideal pesticide would be one which is effective against the target species and has little or no side-effects on human beings, livestock, crop plants, or beneficial and other non-target organisms. Selective chemicals have been of paramount importance in weed control, where damage to crops must be avoided (Fig. 4.1), but most other groups of pesticides have had a broader spectrum of activity. The regulatory authorities have been concerned principally about the toxicity of products and possible hazards to users. Health and safety legislation makes it mandatory in some countries for the appropriate safety precautions to be taken. In many areas, however, especially in the hot and humid tropics, climatic conditions are not conducive to wearing

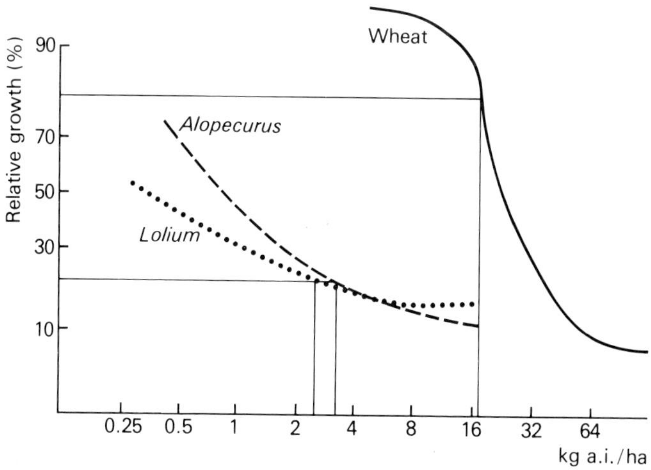

Fig. 4.1 Dose–response curves when screening a post-emergence herbicide to show selectivity between crop (wheat) and weed species (*Lolium* and *Alopecurus*). (From Gerber 1977)

protective clothing, even if users can afford it, so it is not surprising that deaths have occurred following the use of certain highly toxic pesticides.

Further regulations stipulate the time interval before labour can re-enter treated areas or between final treatment of a crop and harvesting to minimize the risk from residues in produce. More recently, greater attention has been given by authorities to the possible effects on wildlife following the realization that some pesticides, notably the organochlorines, can persist for a considerable time, especially in the cooler temperate climates.

The number of pesticides has increased very considerably over the period 1940–80, but the increased cost of providing data for registration has slowed the rate at which new products reach the market. Information on individual pesticides is available from a number of sources, but the most comprehensive catalogue of basic data is *The Pesticide Manual* (Worthing 1983), which is revised at intervals.

A general introductory description of the major types of pesticides is given by Martin (1973), Hassall (1972) and Green, Hartley and West (1977). Lists of approved products are available in many countries with recommendations for control of particular pests, diseases or weeds, or recommendations are given for individual crops. Such data should become more readily accessible with more widespread use of computer-based information retrieval systems. In the UK, the list of products is published under the auspices of the Agricultural Chemicals Approval Scheme.

In the USA, the Environmental Protection Agency (EPA) has to administer legislation concerning pesticides. This includes FIFRA, the Federal Insecticide, Fungicide and Rodenticide Act (Anon. 1981a) covering registration and labelling, the Food, Drug and Cosmetic Act which establishes tolerance limits for shipment in interstate commerce of crops 'adulterated' by pesticide residues and the Federal Environmental Pesticide Control Act. Under the latter Act viruses, bacteria, fungi, sex attractants and insect hormones are subject to the same general testing procedures prior to registration as required for chemical pesticides. Also, pesticides in the 'restricted' category can be applied only by, or under the supervision of, certified applicators. A restriction on some pesticides can be overcome when low concentration granular formulations are recommended or if a product is packaged in an acceptable manner, thus 90 per cent methomyl wettable powder is in water-soluble bags which are inserted directly into a spray tank. Some of the difficulties in making reasoned decisions regarding the regulation of toxic substances are discussed by Findley and Farber (1981) with particular reference to organochlorine pesticides. In an attempt to fill gaps in these statutes, the Toxic Substance Control Act is the first comprehensive legislation governing toxic substances. It covers environmental effects of chemicals and requires industry to obtain appropriate data while the government authority has to prevent unreasonable risks of injury to health and the environment. In implementing this Act, the EPA has so far focused its attention primarily on developing a system to cater for new chemicals but will also re-examine 'old' chemicals.

Similarly, a Committee of Experts on Pesticides has been established by the Council of Europe to study the risk to human health arising from the use of pesticides and make recommendations to ensure their safe use. This committee publishes guidelines for national and other authorities as well as manufacturers which are concerned with pesticide registration.

Most authorities list alphabetically all the products registered for use, so individual users have to decide which is most appropriate for their problem. When the pest is widespread and the potential market for a product is large, several products may be registered for one pest, making it more difficult for the user to decide which to use. For example, twelve different active ingredients were listed for aphid control on beans in the UK in 1980. These included highly toxic insecticides as well less toxic and more selective products.

When minor pests are involved there is less incentive for the agrochemical industry to obtain all the information needed for registration, so few, if any, products may be listed. A user is then left to assume that if a product is recommended against a similar pest on another crop, it may be effective.

Economic factors often dictate that the cheapest in the short term, and most readily available pesticide is used, even though it may not be the most suitable. The selection of the most appropriate pesticide to apply in a pest management programme is not as simple as choosing a single easily obtainable product from the approved list. The aim is not to kill the pest quickly regardless of environmental considerations, but to integrate chemical control as far as possible with other measures. This necessitates a pesticide which is intrinsically more active to the target species than to non-target species (referred to as physiological selectivity) (Graham-Bryce 1977), by improved timing of application and by increasing the proportion of the applied dose which reaches the target organism (referred to as *ecological* and *application selectivity*).

Herbicides

Herbicides have a relatively low toxicity to mammals and an excellent safety record (Fryer 1977), although there are a few exceptions such as the bipyridylium compounds. Selectivity between plant species also has always been of paramount importance in weed control, where damage to crops must be avoided.

The first major selective herbicides were 2,4–D and 4-chloro-2-methylphenoxyacetic acid (MCPA) which continue to be widely used in cereal crops because they are more active against broad-leaved weeds. They are referred to as *hormone weedkillers*, since their biochemical activity on susceptible tissues is similar to the natural plant auxin, 3–indoleacetic acid, which causes the elongation of cells. Thus growth of plants affected by these herbicides is distorted. The selectivity of the compounds is not clearly understood, but some plants have the ability to metabolize them. Another example is the use of simazine for weed control in maize as this herbicide is rapidly converted to a non-toxic compound within maize plants.

Selectivity can be achieved if the applied herbicide is not phytotoxic but has to be converted by the susceptible weed species to a phytotoxic derivative. Several legumes such as peas and clover are resistant to the phenoxybutyric acid derivatives, 2,4–D and 4-(4-chloro-2-methylphenoxy) butanoic acid (MCPB) as they lack the oxidases necessary to convert them into the more active phenoxyacetic acid derivatives, whereas surrounding weeds are controlled. Selective control of grasses in broad-leaved crops is now possible with the development of new post-emergence herbicides such as sethoxydin and fluazifop-butyl. The latter is hydrolysed rapidly after entering the plant, is translocated in both the phloem and xylem and affects meristematic tissue.

These selective herbicides could increase the use of closer-spaced rows of some crops where an appropriate broad-leaf herbicide is available (Stonebridge 1981) provided the chemicals are not antagonistic to each other. Thus in soyabean bentazon or acifluorfen are available as post-emergence broadleaf herbicides which are compatible with fluazifop-butyl. This development opens up the opportunity to use minimum or zero-tillage systems for broadleaved crops in some areas to reduce erosion in a similar manner to that now practised for cereal crops on an increasing scale.

'Chemical ploughing' has been achieved by controlling weeds and crop plant residues using broad-spectrum herbicides, particularly paraquat. Weed control in this way clears the field without ploughing and allows direct drilling of cereals and other crops. Paraquat is only effective when absorbed by plants and spray contacting the soil is inactivated as the chemical is strongly absorbed by clay particles. Where perennial grasses such as common couch grass (*Agropyron repens*) are a problem, glyphosate is better; its effect on foliage is slower, but the herbicide is translocated down into the rhizomes. With reduced disturbance of the soil and retention of a mulch on the surface of the fields, minimum or zero tillage systems are particularly important in areas where heavy rain washes away the topsoil. Broad-spectrum herbicides can be used selectively by directed placement between crop rows or, in the case of wild oats and other tall weeds such as *Sorghum halepense*, volunteer maize and sugar beet bolters, by using weed wipers or recirculating sprayers which achieve selectivity by exploiting the height difference between crop and weeds (Lutman 1980).

More selective herbicides are needed when prolonged use of chemical weed control leads to an increase in the proportion of weeds closely related to the crop. Major weed problems include wild oat, *Avena* spp. in cereals, wild rice in rice and grasses in sorghum. Several post-emergence herbicides have been used to control emerged grasses in cereal crops. Difenzoquat, for example, is active against wild oats while isoproturon is needed to control black grass *Alopecurus myosuroides*. Such herbicides applied early in the autumn on winter weeds give better yield increases. The weed seeds which

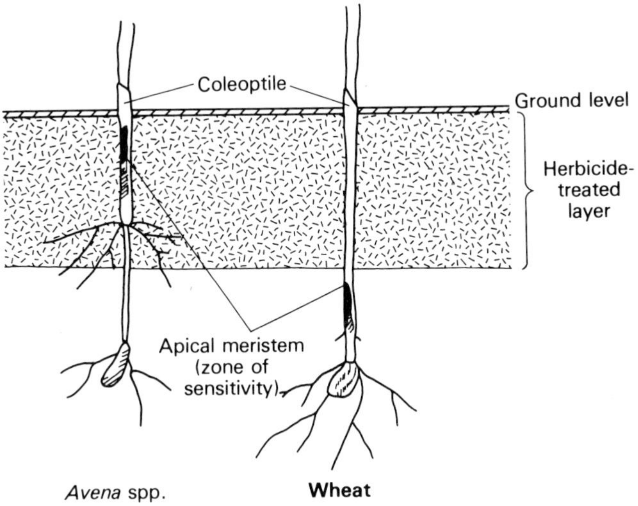

Fig. 4.2 Difference in development of wild oat (*Avena* spp.) and wheat seedlings at 14 days to show that sensitive apical meristem on wheat remains longer below the treated layer of soil. (From Fryer and Makepeace 1977)

germinate late may not be competitive with the crop, but return new seeds to the soil. A sequence of herbicides may be needed therefore during the season with the early treatments to remove serious weed competition. Later treatments are to manage weed population to reduce the risk to subsequent crops.

Another approach which may provide selectivity in the field, is to treat the seeds of the crop before sowing with an antidote or herbicide safener (Gray *et al.* 1982). Subsequent post-emergence application of a herbicide is thus possible while the antidote is effective. The safener NA (1, 8-naphthalic anhydride) as a seed treatment provides protection against herbicides such as perfluidone (Blair and Dean 1976) and DPX 4189 (Parker, Richardson and West 1980). The possibility of mixing the antidote with a herbicide in the spray tank has also been considered (Blair, Parker and Kasasian 1976).

Differential uptake of a soil-applied herbicide is another means of selectivity, by concentrating the chemical near the soil surface where weed seedlings germinate and have their roots, while crop seeds planted deeper contact a lower concentration. The sensitive meristem of wild oat is near the soil surface and more susceptible to a band of tri-allate than wheat sown deeper (Fig. 4.2).

Insecticides

The toxicity of many insecticides to animals other than insect pests has continued to cause great concern among environmentalists. Contamination of non-target areas may occur when particles or spray droplets drift downwind, when volatilization of the pesticide takes place during and after application, by unwarranted disposal of containers near or in streams and ditches, through lateral movement of pesticides in soil and by movement of animals. When animals are under stress, possibly caused by a shortage of food, are migrating or are breeding they are more likely to be vulnerable to pesticides. The major cause of mortality of birds in the UK occurred in the spring when cereal seeds treated with organochlorine insecticides were sown when food supplies were limited. This serious threat to wildlife was removed when the seed treatments were withdrawn and subsequently most uses of the more persistent organochlorine insecticides were restricted (Moore 1977).

The difference in susceptibility of insects and mammals (rats) to a pesticide varies considerably within groups of insecticides (Table 4.1). Examination of these data enables the less hazardous pesticides to be chosen, but from an environmental point of view the problem is more complex. The danger from some of the organochlorines was not caused by their intrinsic toxicity, but the prolonged persistence of residues and their accumulation in food chains.

The synthetic pyrethroid insecticides are considerably less hazardous to apply as far as mammalian toxicity is concerned, but their very broad spectrum of activity against insect species, even at exceedingly low dosages, presents a risk to beneficial and other non-target insects. This can cause resurgence of the pest species, or pest outbreak of another species.

Table 4.1 Comparative toxicity of insecticides to insects and rats based on approximate LD_{50} values expressed as mg/kg. (From Elliott and Graham-Bryce 1977)

Note: Houseflies were used as the test insects except those marked (L) for locusts, (A) for aphids and (MB) for mustard beetles

	Insects	**Rats**	**Ratio toxicity rats/insects**
Organochlorine			
DDT	10	180	18
dieldrin	0.5	40	80
HCH	9(L)	190	21
Organophosphates			
parathion	0.9	3–6	5
diazinon	5(L)	185	37
malathion	18	2,800	156
Carbamates			
aldicarb	10(A)	0.9	0.09
carbaryl	31(L)	540	17
Pyrethroids			
Natural pyrethrin	0.3(MB)	340	1,100
permethrin	0.7(L)	1,500	2,100
deltamethrin	0.03	135	4,500
bioresmethrin	0.2	8,600	43,000

When highly toxic insecticides have to be used, granular formulations are preferred where possible to reduce the hazard to those applying them. Thus soil-applied aldicarb and oxamyl are important not only as systemic insecticides, but principally as nematicides. A toxic insecticide such as mevinphos can be used to clear crops of insects up to one day before harvesting because it is so rapidly degraded.

Now that the environmental impact of a pesticide has to be evaluated to meet local registration requirements, toxicity tests normally include birds, fish and bees. The Japanese quail (*Coturnix corturnix japonica*) is a small bird that has been used, but some tests have been carried out with pheasants and the bobwhite quail. The domestic fowl has not proved very satisfactory for tests on avian toxicity. Rainbow trout or the harlequin fish (*Rasbona heteromorpha*) are commonly used in fish-toxicity assessments. Regulatory authorities have protocols which describe the assessment tests which have to be carried out to meet registration criteria. An overall rating based on these toxicity assessments was used by Metcalf (1975) together with information on mammalian toxicity and persistence to determine which insecticides were most suitable for a pest management programme. Carbaryl, iodofenphos, malathion, methoxychlor, tetrachlorvinphos and trichlorphon are among the safest and most suitable for inclusion in pest management programmes (Table 4.2). Carbaryl has been extensively used on a number of crops, principally to control lepidopteran larvae instead of DDT. A disadvantage of carbaryl, like DDT, is that frequent applications on crops are likely to increase mite infestations. Carbaryl is primarily a contact insecticide, but stomach poisoning can be increased by the addition of molasses and this has been shown against *Heliothis* spp. Carbaryl is rapidly hydrolysed to α naphthol in humid conditions in the soil. An insecticide with a fairly broad spectrum of activity is often preferred because crops are often attacked by more than one pest at the same time. Thus carbaryl is used against different species of bollworms (*Diparopsis, Earias* and in some areas *Heliothis*) and cotton stainers (*Dysdercus* spp.), although it is not suitable for aphids or mite control.

Iodofenphos has been used primarily for protection of stored products and in public hygiene. Malathion has been used in a similar manner, but has also been widely used in agriculture against sucking pests and mites, including many bugs, scale insects and fruitflies. Ultra-low volume formulations of malathion have been applied aerially to control mosquitoes, and mixed with protein hydrolysate bait against *Ceratitis capitata*, the Mediterranean fruitfly. Methoxychlor is similar to DDT, but does not appear to be concentrated in animal fat, and has been recommended for fly control in dairies. Tetrachlorvinphos is highly toxic to houseflies and has been used to a limited extent on some perennial crops. Trichlorphon's activity is attributed to its metabolic conversion to dichlorvos, and is effective against a wide range of insect pests. In South-East Asia, trichlorphon was used to control bagworms in cocoa in preference to dieldrin which has a devastating effect on natural enemies. In the UK trichlorphon and iodofenphos are recommended on

Table 4.2 Examples of toxicity rating of insecticides. (Based on Metcalf 1980, with modifications, and Metcalf and Luckmann 1975).

Insecticide	Mammalian toxicity	Fish	Non-target toxicity Pheasant	Bee	Average	Environmental persistence	Overall rating
A. Most suitable for pest management – overall rating 4–7							
carbaryl	2	1	1	4	2.0	2	6.0
iodofenphos	2	4	1	3	2.7	2	6.7
malathion	2	2	1	4	2.3	1	5.3
methoxychlor	1	3	1	1	1.7	2	4.7
tetrachlorinphos	1	4	1	4	3.0	1	5.0
trichlorphon	2	1	2	1	1.3	1	4.3
B. Suitable for pest management with adequate supervision – overall rating 7–10							
cypermethrin	2	5	1	5	3.7	2	7.7
deltamethrin	3	5	1	5	3.7	3	9.7
diazinon	3	2	5	4	3.7	3	9.7
dicofol	2	1	2	1	1.3	4	7.3
dimethoate	3	1	4	5	3.3	2	8.3
lindane	3	3	2	4	3.0	4	10.0
methyl parathion	5	1	5	5	3.7	1	9.7
mevinphos	5	3	5	4	4.0	1	10.0
permethrin	2	5	1	4	3.3	2	7.3
phosphamidon	4	1	5	3	3.0	2	9.0

C. Only suitable for restricted use in pest management programmes – overall rating 10–13							
aldicarb	5	3	5	5	4.3	3	12.3
aldrin	4	4	4	4	4.0	5	13.0
carbofuran	5	2	5	5	4.0	3	12.0
DDT	3	4	2	2	2.7	5	10.7
demeton-S-methyl	5	2	5	2	3.0	2	10.0
dieldrin	4	4	3	4	3.7	5	12.7
disulfoton	5	3	5	2	3.3	3	11.3
endosulfan	4	4	4	4	4.0	3	11.0
phorate	5	4	5	2	3.7	3	11.7
D. Not suitable for pest management – overall rating > 13.							
endrin	5	5	5	2	4	5	14.0

brussels sprout crops to protect against the larvae of cabbage rootfly and other insects. The botanical insecticides, natural pyrethrins, nicotine and ryania generally have a low persistence. Pyrethrins are used mainly in public health work, while nicotine has been used in glasshouses. Ryania has been used against codling moth in an integrated control programme.

Very few intrinsically selective insecticides have been marketed because of the high cost of evaluating new products to meet safety standards. These costs cannot be recovered if a selective chemical has only a limited market. Some specific aphicides have been developed. Pirimicarb is relatively very safe to beneficial species except *Trichogramma* and syrphids, and has replaced menazon which was as toxic to bees as DDT.

With selective aphicides, the dosage needs to be carefully selected if used in an integrated control programme. Unless a high level of control is needed rapidly to check the spread of a virus, too great a mortality as a result of a spray application will leave insufficient food for predators. In consequence, any predators that are present will emigrate or starve and others are not attracted into the crop, so none are available if a later immigration of aphids occurs.

The pesticide diflubenzuron, originally introduced as a fungicide, interferes with the deposition of chitin at moulting (Mulder and Gijswijt 1973), and so is an important insect growth regulator (IGR) to control lepidoptera larvae (Grosscurt 1978). Unfortunately, the larval stages of some beneficial insects such as coccinellid larvae are also affected. It is relatively safe to adult predators and parasites and to all stages of predaceous phytoseid mites. Diflubenzuron is recommended for codling, some tortricids and winter moths in apple orchards, but these pests have to be controlled at the egg/first instar stage, before the individuals become protected within the plant.

Recently the insecticide buprofezin, a thiadiazin (Kanno *et al.* 1981) has been shown to be effective against the nymphs of the brown planthopper *Nilaparvata lugens*, but is of low toxicity to its natural enemies (Kajihara *et al.* 1982). In the field it will reduce a brown planthopper population, and for a quick knockdown can be mixed with a carbamate insecticide.

The juvenile hormone mimicking insecticide methoprene is now commercially available as a mosquito larvicide, but due to its rapid degradation, more persistent slow release and micro-encapsulated formulations have had to be developed. The possibility of using an anti-hormone mimic to produce precocious adult insects is also under investigation.

The list of insecticides useful in a pest management programme can be widened if care is taken to restrict the area of application and dosage to minimize undesirable side effects. The use of synthetic pyrethroids in agriculture has expanded enormously following the development of photostable compounds with a broad spectrum of activity. They have been used as alternatives to DDT, since they are readily metabolized in mammals and do not accumulate in food chains, though they are extremely toxic to fish. This hazard can be reduced by avoiding contamination of waterways and by using very low dosages effective against insects. Activity against mites is poor although some pyrethroids are said to have some acaricidal activity. Other synthetic insecticides should be avoided as far as possible, although individual chemicals may be needed for specific pest situations. Endosulfan has been successfully used in sequential aerosol sprays over large areas of Africa against tsetse flies (*Glossina* spp.) as it is effective at extremely low dosages. Elsewhere in Africa endosulfan has been widely used on cotton.

Most insecticides in group C (Table 4.2) are extremely hazardous, so are locally applied alongside the seed as granules and incorporated in the soil. In contrast, DDT has a good safety record and remains an effective insecticide against a number of important pests including mosquito vectors of malaria. In general the persistent organochlorines are the least suitable insecticide for any pest management programme and alternatives to them will be used on an increasing scale. Spray deposits are exposed to rainwashing and degradation due to ultra-violet light, so the persistence of the organochlorines used to be considered an important attribute. Some persistence is needed as infestations may continue over a period of days or weeks. Busy farmers would prefer long intervals between treatments, but growth of plants increases the area of untreated foliage so rapidly in some situations that treatments are needed at least at weekly intervals over the critical period irrespective of how persistent the pesticide is. This applies not only to insect pests; thus apple powdery mildew can be very difficult to control as young, untreated leaves are easily infected. There is a compromise to be found between persistence for effective control and long-term contamination of the environment.

'Acaricides'

Some chemicals are highly toxic to mites and ticks, but are relatively harmless to insects, although chlordimeform used as a mite ovicide is also effective against eggs of lepidopterans. Amitraz is similarly effective against phytophagous mites and some insects. Some pesticides with acaricidal properties such as binapacryl are also very effective fungicides, so there is a risk that when some are used as fungicides, they may have a detrimental effect on predaceous mites. Tetradifon is a non-systemic acaricide, toxic to the eggs and all non-adult stages of phytophagous mites, while dicofol is active against all stages except overwintering eggs.

As selection for mites resistant to pesticides can be quite rapid, rotation of acaricides around different areas of a country has been recommended in order to reduce the exposure of populations to a particular compound (Duncombe 1973).

Fungicides

Choice of fungicide will depend to a large extent on whether control of surface spores is the aim or whether the mycelium of the fungus within the plants is the target. Most of the older fungicides including copper compounds such as copper oxychloride, sulphur and captan, are applied as protective sprays to inhibit germination of spores or penetration of plant tissues by the hyphae. Such chemicals have little or no effect on fungal pathogens within plant tissues, so there has been widespread use of systemic fungicides such as benomyl, carbendazim, triadimefon and tridemorph which have both protective and curative action. Triadimefon is recommended on cereals to control mildews and rusts. Unfortunately, resistance to systemic fungicides appears to occur very rapidly so alternative use of systemic and protectant fungicides has been suggested to try to delay the onset of resistance (Spencer 1972). The majority of fungicides are relatively safe pesticides to mammals but some have significant effects on non-target organisms; thus benomyl residues in orchards can reduce earthworm populations (Cooke, Herrington and Morgan 1974) and excessive applications of copper fungicides year on year may also result in a reduction in the number of soil-inhabiting organisms due to an accumulation of copper. There is the possibility that the food supply of some beneficial species especially mycophagous predators is reduced by fungicide sprays and this can result in an increased insect pest infestation.

Pesticides and bees

Bees are susceptible to most insecticides (Table 4.3) and some fungicides including benomyl, so choice of chemical in areas foraged by bees may be determined by the persistence of spray deposits. Time of application to avoid direct treatment of flowering crops is clearly important as is the period during the day when the insecticide is applied. There is less risk to bees if spraying is carried out in the late afternoon or evening, when they are not active. Micro-encapsulated formulations are not suitable, since bees tend to collect small particles resembling pollen. Granular application is generally safer unless a systemic insecticide is involved, in which case it is likely to be taken up into the nectar. Treatment of crops such as oilseed rape is a particularly difficult problem. A review of the impact of pesticide on bees is given by Crane and Walker (1983).

Interaction of pesticides with predators and parasites

Consideration has been given so far to non-target organisms which may or may not enter the treated area. With integrated control, the effect of pesticides on beneficial organisms, the predators and parasites, active within a treated area, must also be determined. Standardized assessment procedures are being developed to assess the initial toxicity of pesticides to

Table 4.3 Insecticide toxicity to worker honey-bees. (Acute contact and oral toxicity, adapted from Stevenson, Needham and Walker 1978) Walker 1978)

	Acute contact	Oral toxicity (LD_{50} mg/bee)
Insecticides less toxic to bees		
DDT	3.9	3.70
dicofol	> 50.0	> 10.00
diflubenzuron	> 30.0	> 30.00
endosulfan	7.1	6.90
menazon	4.3	0.46
pirimicarb	> 54.0	3.20
Insecticides highly toxic to bees		
carbaryl	1.30	0.14
chlorpyrifos	0.06	0.25
deltamethrin	0.05	0.08
demeton-S-methyl	0.26	0.21
dimethoate	0.12	0.15
fenitrothion	0.02	0.02
malathion	0.27	0.38
permethrin	0.11	0.28
triazophos	0.06	0.07
Other pesticides		
benomyl	> 10.0	> 10.0
captan	> 10.0	—
MCPA	> 100.0	> 10.0
triadimefon	> 25.0	> 25.0

predators and parasites by the International Organization for Biological Control of Noxious Animals and Plants (IOBC). Test procedures have been published by Hassan (1977), Franz, Hassan and Bogenschutz (1976), Plattner (1975), Viggiani and Tranfuglia (1978), Franz *et al.* (1980) and Suter (1978). Hislop and Prokopy (1981) have assessed the influence of pesticides on the predator *Amblyseius fallacis* under both laboratory and field conditions. Ledieu (1979) developed a technique for assessing the effects of various pesticides against adult *Encarsia formosa*, an important parasite of whiteflies (Fig. 4.3).

The most important aspect of such studies, apart from measurement of direct mortality, is whether those which survive sub-lethal doses are able to search for a host and oviposit. The fungicide quinomethioate was found to be more harmful to *Encarsia* in glasshouse tests where a possible repellant effect prevented the adult parasite ovipositing into treated whitefly scales (Ledieu 1979). Some insecticides such as the pyrethroids and DDT have irritant effects which may prevent some insects remaining on deposits long enough to pick up a lethal dose. Sub-lethal deposits may also prevent feeding, so all the available information on the effects of each chemical on natural enemies needs to be collated when attempting to integrate chemical and biological control.

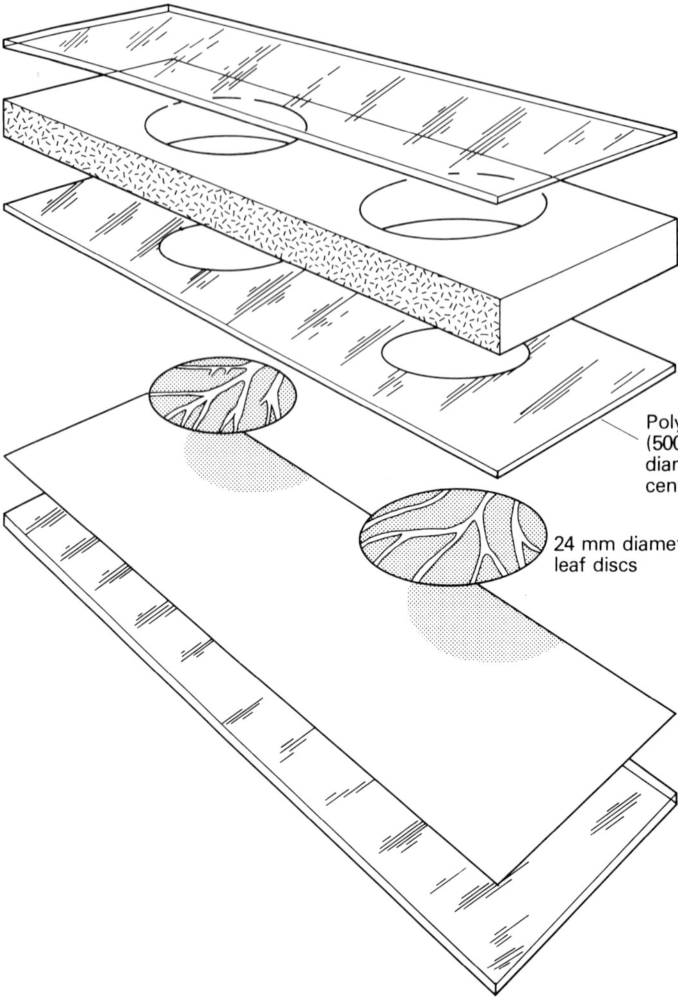

Fig. 4.3 Apparatus for laboratory screening of pesticide against adult *Encarsia formosa* (From Ledieu 1979)

A typical summary of this type of information is shown in Table 4.4; here it is designed to guide the user in his choice of pesticide in glasshouse crops where *Phytoseiulus* is released to control mites and *Encarsia* to control whiteflies.

The impact of a pesticide on the beneficial capacity of natural enemies needs to be considered more carefully and the safest and least persistent chemical should be chosen. For this reason insecticides which leave toxic deposits on coffee leaves for more than two or three days are rejected (Bardner 1978).

Laboratory tests may indicate that a pesticide is harmful, but the harmful effects may be lessened in the field by localizing the treatment or by careful

Table 4.4 Example of information on insecticides suitable for the control of aphids, red spider mites, whiteflies and leaf miners on tomatoes in relation to their effect on natural enemies.

Compound	Application rate	Minimum harvest interval	Active against: Aphids	Active against: Red spider mite	Active against: Whitefly	Active against: Leaf miner	Effect (where known) on: Phytoseiulus Eggs	Phytoseiulus Adults	Encarsia Pupae	Encarsia Adults
HV Sprays	g/100 litres	*days*								
cyhexatin□	25	2	—	Yes	—	—		H	S	S
deltamethrin	1.75	0	Yes	—	Yes°	—	H	H	H	H
demeton-S-methyl$^{+\triangle}$	22	21	Yes*	Yes*	—	—	H	H	S	H7
diazinon	16	14	Yes*	Yes*	Yes$^{*\phi}$	Yes*		H	H	H7
dichlorvos	50	1	Yes	Yes	Yes$^{*\phi}$	—		H		H2
dicofol	11.5	2	—	Yes*	—	—		H	S	S
dicofol/tetradifon	40/12.5	2	—	Yes*	—	—	I		I	I
dimethoate^{+}	30	7	Yes*	Yes*	—	—	H	H	H	H > 7
endosulfan$^{\int}$	88	14	—	—	Yes°	—		H4		
gamma-HCH (lindane)	12.5	14	Yes	—	Yes	Yes*	H	H < 4	H	I < 4
heptenophos^{+}	41	1	Yes	—	—	—		H < 3	S	I < 4
malathion	114	1	Yes*	—	Yes$^{*\phi}$	Yes	H	H10	H	H10
oxydemeton-methyl$^{+\triangle}$	22	21	Yes*	Yes*	—	—		H		
parathion$^{\S\triangle}$	10	28	Yes*	Yes*	—	Yes	H	H14	H14	H
permethrin	12.5	0	Yes	—	Yes°	Yes		H	S	H
petroleum emulsion	‡	7	—	Yes	—	—	S	S	S	I
pirimicarb$^{\triangle}$	25	3	Yes	—	—	—	S		I	
pyrethrum/resmethrin$^{\triangle}$	‡	0	Yes	Yes	Yes°	—	H	H < 4		H < 4
tetradifon$^{\triangle}$	12.5	0	—	Yes*	—	—	S	S	S	S

* — Possibility of encountering resistant insects
ϕ — Kills adults only O — kills adults and larvae
H — Harmful (H >7 — harmfulness persists for more than 7 days)
I — Intermediate
S — Safe
$\wp$ — *Liriomyza trifolii*; use 6g/100 litres
‡ — See product label
□ — Avoid spraying fruit < 1 cm diameter
\+ — Systemic compound
△ — Young plants may be treated
§ — Maximum of 3 applications/season
∫ — Maximum of 6 applications/season

timing, in which case the beneficial species is less exposed to the spray deposits. Obviously, no pesticide should be applied if natural enemies are being released and have not yet become established. Apart from direct effects on mortality, pesticides can be harmful by interfering with fecundity or searching behaviour, thus reducing the ability of a natural enemy to exert biological control.

When particular pesticides have been used in a given locality for some length of time, natural enemies may adapt and become behaviourally or biochemically tolerant. Croft (1979) pointed out that where organophosphate insecticides have been used in apple orchards for many years, growers benefit from there being natural enemies tolerant or resistant to OP compounds. If a change were made to alternative chemicals such as the synthetic pyrethroids, it would take between five and fifteen years before certain beneficial species could re-adapt, if at all, to the new pesticides and reassert their biological control potential. A broad spectrum insecticide can also be applied selectively, thus in coffee a pyrethrin 'knockdown' spray causes the giant coffee looper and its natural enemies to fall from the foliage. The loopers, which have to climb back up the tree, are killed by DDT in a band painted around the base of the tree trunk, while the adult natural enemies fly back and avoid the DDT (Wheatley, 1963). (See also page 131).

The influence of herbicides on weeds and insect natural enemies

The harmful effects of weeds competing for nutrients and as reservoirs of viruses (Heathcote 1970), nematodes (Franklin 1970) insects (van Emden 1970) and other pests, are generally considered to far outweigh any small benefit that they may confer, e.g. as a source of food for natural enemies (Zandstra and Motooka 1978). In parts of the USA, spider mites on weeds in apple orchards provide prey for an important predaceous mite, allowing it to survive early in the season (Croft 1975). However, the presence or absence of weeds can have quite contrasting effects on some predators; thus in East Africa control of the coreid *Pseudotheraptus wayi* by *Oecophylla longinoda* is better when there is weed cover (Way 1953), whereas in the Solomon islands, elimination of weeds around each palm tree with paraquat favours *O. smaragdina* and allows greater control of another coreid *Amblypelta cocophaga* (Stapley 1973). Colonization of plants by some insect pests can be decreased when weeds act as a camouflage and reduce the colour contrast between green plants and bare soil (Smith 1969, 1976). On the other hand, when herbicides are used plants can be sown more closely or evenly to ensure the canopy covers the ground earlier to limit the period of attractiveness (Way and Cammell 1981).

Herbicides apparently have relatively minor direct effects on the soil fauna (Edwards 1970) compared with those caused by mechanical weed control, which can decrease survival of some overwintering natural enemies (Way, Murdie and Galley 1969). Apart from some *Coleoptera* and *Diptera* spp., soil organisms and some pests such as slugs, were more abundant in herbicide-treated direct-drilled plots than in conventionally cultivated soil (Edwards 1975).

Some herbicides, notably 2,4-D, can be toxic to certain beneficial insects and may be responsible for increased pest incidence on cereals (Adams and Drew 1965) and the decrease in diversity of organisms in treated fields (Potts and Vickerman 1974; Vickerman 1974), although changes in abundance of cereal aphids may be associated with other changes in crop production technology.

Way and Cammell (1981) have suggested that where weeds are scarce, pests may become more adapted to crops and this will increase selection for resistance to pesticides on treated crops. In general. too little is known about the importance of weeds or the effect of herbicides in the context of an integrated control programme.

5

What dosage?

Organisms differ in their susceptibility to a given pesticide; biological assessments or bioassays are therefore used to determine the dosage appropriate for individual pest species. Bioassays are strictly experiments involving the use of living organisms to determine the amount of a chemical, in this case a pesticide, in a substance. However, the term 'bioassay' is used in a wider sense to cover all experiments where the potency of a pesticide, drug or, indeed, any compound having a measurable effect on living organisms, is measured using standardized laboratory populations of suitable species. Pesticide companies have large programmes designed to screen compounds for pesticidal activity, and these tests are followed by detailed toxicological studies of promising chemicals to determine whether or not the potential pesticide can be used safely. Only a small number of species which can be maintained easily in the laboratory – commonly a housefly, cockroach, lepidopterous larva and aphids – are used in the initial evaluation to detect insecticidal activity. Similarly, herbicides are screened against a limited number of broad-leaved and grass weeds and representative crops to assess selectivity (Holly 1982). Such tests are used by manufacturers to decide which of the many thousands of compounds tested each year should be developed. Only pesticides effective against major pest species will have a sufficiently large market to recoup the enormous financial investment in toxicological testing and other development work.

Those in the field who have to decide which pesticide to use in a pest management programme rely very much on the information submitted by the chemical companies to the registration and approval authorities. More detailed evaluation of a range of pesticides may be made independently by government or university scientists. Test procedures are often similar to the screening programmes used by the commercial companies, but there is greater scope to adapt basic bioassay techniques to obtain more detailed information relevant to particular pest problems. In these circumstances, choice of a bioassay technique will depend to some extent on an initial study of the life cycle of a pest to determine which stage is more accessible to control. As an example, internal feeders of plants such as cotton bollworms, are exposed to spray droplets as adults, eggs and first instar larvae. Later larval instars are protected within the boll, so a pesticide is needed which will affect the adults and reduce oviposition, be ovicidal or control the larvae before they penetrate into the plant tissues. Different bioassay techniques are needed in each case, thus an aerosol spray will be needed to optimize pick-up by a flying insect, whereas a residual film on a plant surface will be more suitable for studying ovicidal or larvicidal activity. Spray coverage is probably less critical where movement of the insect is liable to increase pick-up. Similar considerations apply in the evaluation of other pesticides, thus for herbicidal activity, tests are devised to assess activity against young seedlings emerging through the soil for pre-emergence application and their effect on established plants for post-emergence treatment. Herbicides also need to be evaluated at different growth stages of both weeds and crops to ascertain the optimum time for treatment with maximum safety to the crop.

Individual insects or plants may be treated each with a dose, a defined quantity of toxicant at one time or in a given period of time, but application of a precise dose is not always possible even under laboratory conditions and it is more usual to apply the toxicant to a surface or group of individuals and this is referred to as the *dosage applied*.

Most assessments of pesticides involve exposure to a single dose or dosage and thus measure the effect of acute poisoning, but in practice an organism may pick-up repeated small doses over a longer period of time. Chronic poisoning is very much more difficult to measure as effects will undoubtedly be influenced by the size of individual doses, the intervals between exposure and the rate of which the pesticide is absorbed, metabolized and excreted.

Chronic toxicity tests are particularly important where pesticides are persistent in the environment thus increasing the risk of continued exposure to deposits. In pest management, the effect of sub-lethal dosages on the behaviour as well as mortality of both pests and natural enemies needs to be examined.

A brief description of some bioassay techniques is given here with particular reference to insecticides, although similar spraying techniques are used to treat surfaces such as plants and the soil when testing fungicides and herbicides. Shepherd (1958) and Busvine (1971) should be consulted for a critical review of a wide range of methods of testing insecticides. Some of the techniques have been extensively used with dilute sprays at high volume, but when newer techniques adapted for ultra-low volumes are used, spray concentration can be a critical factor. Choice of formulation is also very important, some compounds being more active if a suitable adjuvant is included; e.g. the activity of difenzoquat is dependent on the concentration of surfactant in the spray (Merritt 1982a).

Factors which affect biological assays

Several factors are likely to influence the apparent potency of an insecticide, so standardization of test procedures is important, especially when results of tests undertaken at different times need to be compared. These factors are: 1, intrinsic differences between various types of insects; 2, fluctuations in susceptibility caused by extrinsic changes in the environment; 3, experimental factors, including differences between formulations and method of application.

Intrinsic factors

Species specificity

In the last chapter reference was made to the differences in susceptibility and the need to seek pesticides active against pests and if possible leave beneficial insects unharmed. As new compounds are screened against relatively few representative species, biological assays are subsequently carried out with promising materials to ensure that an appropriate chemical is applied in the field. Species specificity may be due more to differences in uptake of a pesticide and penetration, than to metabolic differences, degradation, or intrinsic toxicity. Busvine (1971) quotes the classic example of the variation in sensitivity of certain lepidopterous larvae to the insecticide rotenone (Table 5.1), the silkworm being nearly 2000 times more susceptible to this compound when topically applied.

Table 5.1 Sensitivity of certain lepidopterous larvae to the insecticide rotenone

Insect	LD_{50} (mg/g)	Reference
Silkworm *Bombyx mori*	0.003	Shepherd and Campbell 1932
Bollworm *Heliothis zea*	> 0.49	Hansberry and Richardson 1936
Armyworm *Spodoptera eridania*	> 5.0	Woke 1938

However, when rotenone is injected into *Spodoptera* larvae, it is similarly toxic, so in this case the greater tolerance is apparently due to poor penetration. More recently carbaryl was shown to be more effective than DDT against *Diparopsis castanea*, but the reverse was true with *Heliothis armigera* (Matthews 1966). Quite possibly, important pesticides may be missed if they are not effective in the initial commercial screens.

Stage specificity and age

Changes in the anatomy, physiology, size and behaviour of the individual instars can influence the susceptibility of a pest. Similarly in plants, there may be particular growth stages which are more susceptible to herbicides. Plants under stress, possibly due to a nutrient deficiency, or damaged, may be more susceptible to a herbicide. Some herbicides are only effective when the weeds are actively growing.

In topical application studies, Gast (1959) showed that some insects increase their tolerance to certain insecticides as they aged; thus the LD_{50} (mg/g) for *Heliothis zea* larvae increased thousandfold as larvae grew from 0.1 to 0.5 g weight. Such a large increase may not always occur, e.g. there is almost uniform tolerance in *Spodoptera eridania* treated with mevinphos, irrespective of size (Fig. 5.1).

When considerable changes in tolerance occur, field applications need to be accurately timed to control the most susceptible stage, normally the first instar. Apart from the very much greater and uneconomic dosage needed for later instars, control at this stage is too late to prevent damage, and the residues of excessive amounts of insecticide are likely to persist longer in food and the environment and have adverse effects on natural enemies. The susceptibility may also vary during the life of an instar; in *Diataraxia oleracea*, for example, tolerance increased four or five times to the middle of the instar, then fell to about double the initial LD_{50} value before the next moult (Mukerjea 1953).

Sex

Males of many species are more susceptible to insecticides than the females. In some cases these differences may be related to the respective size of the sexes, but even when the dosages are corrected for body weight, the female often has greater tolerance. Bioassays should therefore be carried out with either all of one sex or an equal proportion of both in each test.

Size

In pharmacology, it is usually accepted that a larger animal requires a bigger dose of a drug than a smaller one, though in proportion to body weight the amount of toxicant needed is often reduced. However, in practice the recommended dose will have to take account of possible reduced uptake or increased metabolism and that only a small proportion reaches the site of action.

Extrinsic factors

Temperature

Most insecticides have a positive temperature coefficient, i.e. they become more toxic at higher temperatures, but some, like DDT, are said to have a negative temperature coefficient. If such a insecticide is picked-up by insects which are then transferred to a clean surface, more of the insects will be killed with a given dosage at a lower temperature. This effect may be

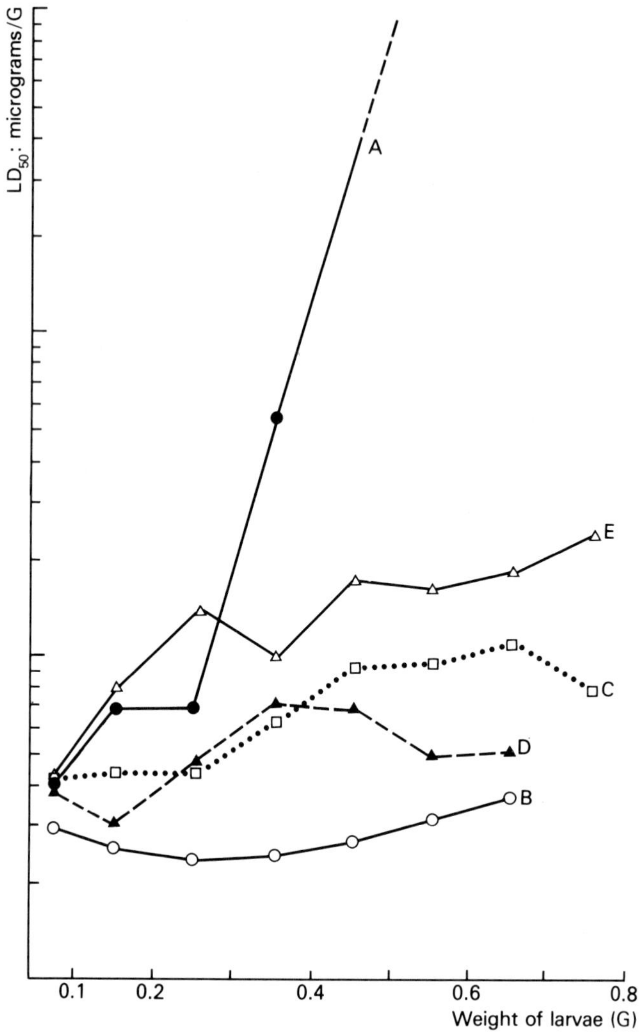

Fig. 5.1 Relations between weights of lepidopterous larvae and lethal doses of contact poisons for unit weight
A *Heliothis zea* and DDT
B *H. zea* and mevinphos
C *Spodoptera eridania* and DDT
D *S. eridamia* and mevinphos
E *Manduca sexta* and endrin
(From Gast 1959; Busvine 1971).
Note: Actual doses are not shown on log-scale since they differ in orders of magnitude from the various insecticides

masked if the insects are retained on the treated surface throughout the test period since insects are usually more active at a higher temperature and this may result in more of the deposit being picked-up by the tarsi. Activity will decrease if temperatures are unfavourably high. The main factor to consider is that in laboratory assays, the temperature should be kept constant to allow comparison of data from different occasions. The selected temperature should be similar to that likely to occur when the pest has to be controlled in the field.

Humidity

This is another factor which should be standardized as far as possible. Some results have indicated greater mortality at higher humidities (Pradhan 1949). Humidity will fluctuate quite considerably in the field and can have important effects on the survival of young lepidopterous larvae, e.g. the survival of bollworm larvae, which hatch early in the morning when the humidity is high, decreases when they are placed on foliage in the laboratory under drier conditions. Provision of a constant humidity is very expensive and, although humidity can be controlled in small enclosed containers such as glass desiccators with the aid of saturated solutions of a suitable salt (Stokes and Robertson 1949), it is inappropriate if the insecticide has any vapour action.

Food

Tolerance to an insecticide may be affected by food in two ways. The quantity and quality of food will affect the growth of an insect prior to treatment so greater size and ability to metabolize the insecticide may increase its tolerance. A difference to tolerance between recently-fed and starved individuals may be apparent, as the presence of food may affect the bioassay. Nevertheless, food has to be provided after treatment to prevent unduly high control mortalities, particularly in long-term tests.

Population density

Some insects such as *Heliothis* larvae are cannabilistic so have to be kept separate in culture and after treatment. When more insects are retained on a treated surface, mutual disturbance may increase activity and reduce tolerance to the insecticide. Although the number of individuals exposed in each treatment is a factor in the calculation of LD_{50} values, the number should be as far as possible the same in each treatment.

Illumination

Some insects are more active at a particular time of the day, thus houseflies which are more active by day are more sensitive to DDT residues in light than darkness (Kettle 1949). Diurnal tolerance rhythms have also been found in many species, so for repeatable results in critical laboratory bioassays test insects should be reared in a defined photoperiod and the insecticide treatments carried out at the same time of day. Biological assays with *Diparopsis castanea* first instar larvae were usually carried out from 5 to 10 a.m. when the eggs hatched (Matthews 1966). Tests to assess aerosols against nocturnal moths would require a very low light intensity equivalent to dusk or moonlit nights. Some plants are more sensitive to herbicides applied in bright light.

Experimental factors

The design of the experiments in terms of the number of replicates and number of individuals in each test, and the method and time of determining the response, should be standardized if results from different experiments are to be compared. An untreated control is essential to check whether the method of handling influences the response to treatment. Analysis of the data can be adjusted to allow for some mortality in the untreated control, but every effort should be made to improve the method used in the bioassay to keep the natural mortality to a minimum. Often one or more extra 'reference' treatments are also needed. If a particular diluent is used, then one treatment should include the diluent without insecticide. Sometimes a 'standard' insecticide. The susceptibility to the 'standard' insecticide may vary on different dates with other batches of insects, but the results achieved with the test insecticide can be compared with the standard.

Great care must be exercised when handling insects before, during and after treatment. Some delicate insects such as aphids are particularly difficult to move, so considerable skill and experience is needed to avoid mortality through handling. When more than one person is involved in treating insects, and assessing the effects of the treatment, standardized techniques must be adopted so that personal idiosyncrasies do not alter the results. Standardization is necessary in any work that relies on an assessment. Even mortality is not always easy to judge. Various criteria have been adopted e.g. mosquito larvae, 'dead' and 'alive' may be separated by a shutter after swimming from light and 'live' beetles classified as those that escape from a pile in the centre of an arena. Sometimes insects appear dead yet recover some hours later and continue to develop normally. This 'knock down' effect is very noticeable with pyrethrins.

In relation to pest management, bioassays involving residual spray deposits or direct spraying of insects are a closer approach to natural conditions and to what happens in the field, but topical application tests give a more precise dose to measure intrinsic toxicity and their results are generally more reproducible. Choice of the test method will depend essentially on the objective of the experiment. Assessment of intrinsic toxicity may be needed to determine the most appropriate chemicals. Further tests will then establish appropriate formulations and dosage rates for field use.

Biological assays

Residual films

A simple residual film insecticide assay can be performed by preparing an insecticide solution from which serial dilutions in a suitable solvent, often by factors of 2, 4, 8 or 16, are made. The solvent may contain one part of an involatile solvent, e.g. Risella 17 oil, and four parts of a volatile solvent such as petroleum ether. Small volumes (0.5 ml) of each of the dilutions and plain solvent as control are pipetted onto a series of filter papers on a non-absorbent surface. There should be at least four replicates of each treatment. Each of the papers is marked with the concentration of insecticide and the time. Later, when the solvent has volatilized and blown away, the insects are caged over the deposit. Open metal rings, treated with Fluon if necessary, are suitable for retaining many insects on the treated surface

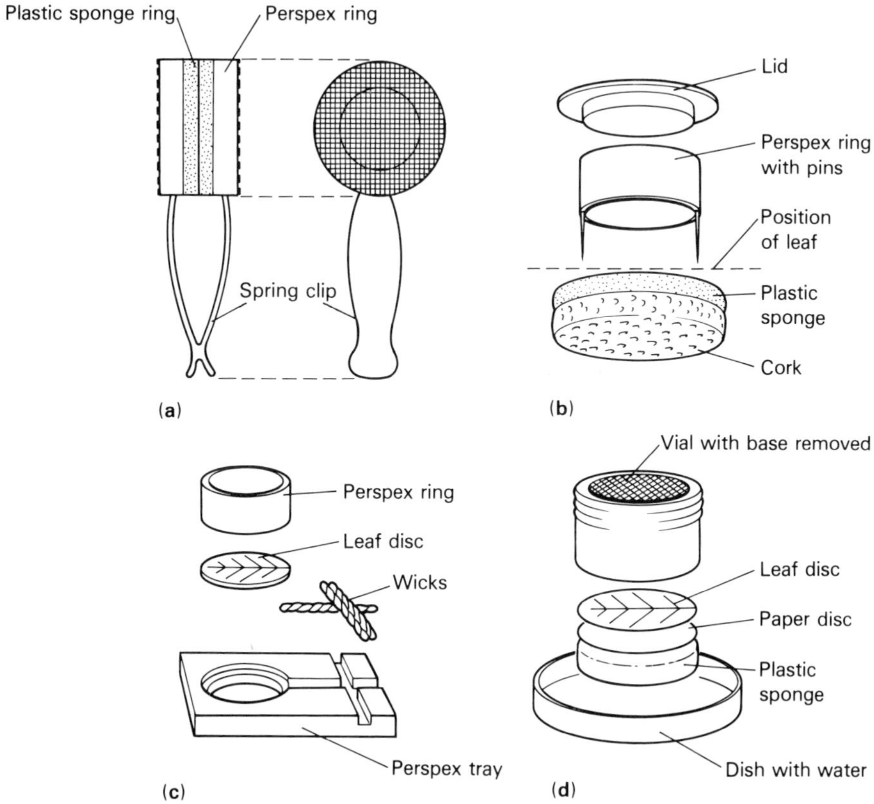

Fig. 5.2 Devices for keeping insects on leaf surfaces. (From Busvine 1971)
(a) Clip-on cage
(b) Pin-on cage
(c) Leaf disc unit
(d) Leaf disc unit

(Fig. 5.2). Various types of restraining devices can be used, including petri dishes, if the lid is modified to allow sufficent airflow to eliminate a fumigant action. The mortality is noted at intervals, and the test should be continued until the endpoint is reached. However, from a practical point of view, assays are often terminated after 24 hr, even though important effects may not take place until later. For example, some insecticides such as diflubenzuron will affect the next moult. If mortality is high after a short exposure period, the assay could be repeated with the insects on the treated surface for a short period before being transferred to a clean container.

The World Health Organization use a modification of the assay described above to monitor insecticide resistance in mosquitoes, using standardized impregnated papers. The mosquitoes are introduced into a plastic tube 12.5 × 4.5 cm diameter with a metal gauze sealing one end and a sliding shield at the other. An exposure chamber, a second tube lined with treated paper,

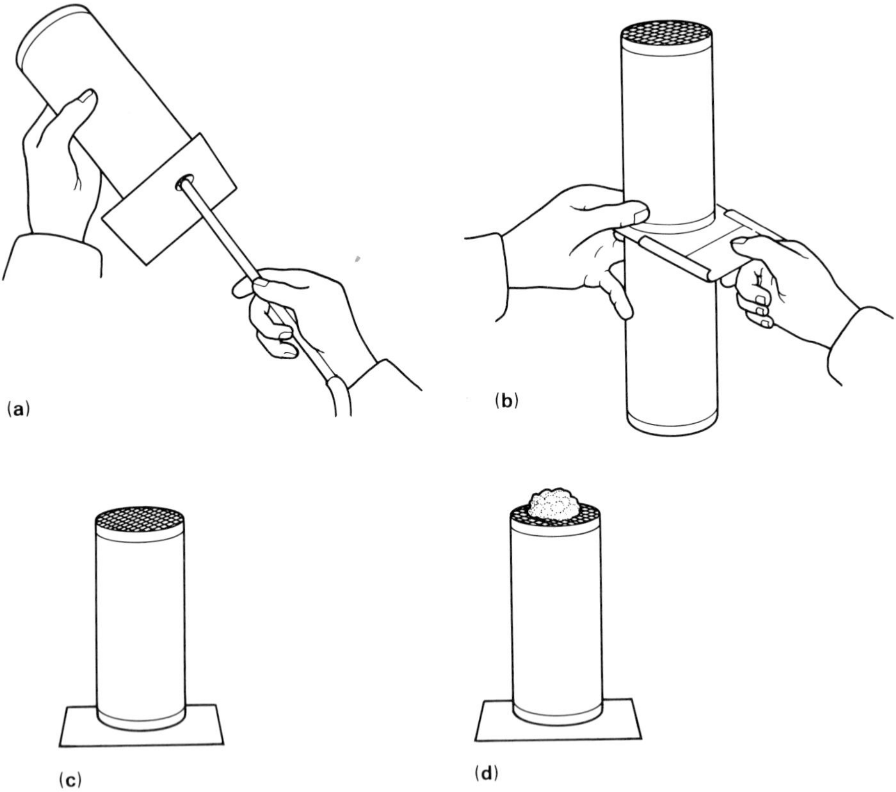

Fig. 5.3 Method for assessing resistance in adult mosquitoes used by WHO
(a) Introducing mosquitoes into the pre-test holding tube
(b) Transferring mosquitoes from the upper, holding tube to the lower, exposure tube
(c) Exposure
(d) Post-test recovery period
(From Busvine 1971)

is screwed on the other side of the slide, which is opened to transfer the insects (Fig. 5.3). An exposure time of 1 min is normally used and the insects are then returned to the untreated container for 24 hr. The same system has been adapted for a number of insects.

Sprayed surfaces

Spray towers have been designed as precision instruments for laboratory research on 'toxicological' problems. A spray tower has to be designed to minimize turbulence, reduce the amount of spray deposited on the sides, and obtain even deposits consistently. Other apparatus has been developed to simulate field-spraying treatments.

Potter tower

Potter (1941, 1952) used a twin-fluid nozzle carefully centred at the top of an open-ended metal tube to spray vertically down on to a horizontal plate 125 mm below the lower flange of the tube (Fig. 5.4). The tower is tapered at the top to create turbulence which evens-up the stream of spray. Various improvements in the nozzle design (Hewlett 1946) and modifications to the tower (Hewlett 1947) have been incorporated into the original design. Spray droplets were often electrostatically charged so to avoid affecting deposition, Potter used an earthed metal tube instead of a glass one. Deposits ranging from 1.76 to 5.7 mg/cm were obtained over 64 cm^2 when 5 ml of different liquids was sprayed (Potter 1952).

As the spray volume corresponds approximately to 200 litres/ha, more recent modifications have aimed to reduce the volume of spray to the order of 5 litres/ha and have incorporated a microsyringe with an electrical drive system similar to that described by Arnold (1965). An alternative system devised by Johnstone (1970) interposed metal plates drilled with holes between the spray and the target to act as screens. Plates varying from 0.9 to 0.2 mm thick were drilled with holes from 0.4 to 64/cm^2. Johnstone studied various factors affecting the density and form of the spray deposit, and showed that the ratio of the depth of hole and hole diameter was important. Fewer droplets impinged on the walls of the holes when the depth was small relative to the diameter of the hole. Increasing the distance between the screen and the spray target increased the scatter of droplets below the holes. The most uniform deposit was obtained when the nozzle was operated at a low pressure (0.5 kg/cm) and the spray volume was restricted to avoid coalescence of droplets below the holes.

Koopman and Balk (1973) moved a twin-fluid nozzle manually over a chamber containing several plants. If specialized equipment is not available, simple hand-held sprayers can be used, but it is difficult to reproduce treatments with accuracy.

Fig. 5.4 Potter tower. (Photograph, Burkard Manufacturing Company)

Spinning disc sprayers

A spinning disc can be used to produce droplets of different size, by adjusting disc speed and flow rate. Kennedy, Ainsworth and Toms (1948), Parr and Busvine (1948), Ennis and Williamson (1963), Byass and Charlton (1968), Lake and Taylor (1974) and McKinley *et al*. (1972) are among those who have used this method for spraying surfaces. The speed of a small disc driven by an electric motor can be adjusted by a voltage control, but the actual rotational speed should be checked with a strobe light or phototachometer, since an increase in flow rate can slow the disc (Fig. 5.5). Volume of spray fed to the disc can be controlled by an electrically-driven syringe. The disc is usually shrouded so that spray is emitted through a slit which can be adjusted so that droplets fall as a fan-shaped curtain on to the target

Fig. 5.5 Spinning disc sprayer.
ULV/CDA track sprayer used in the bioassay of spray droplet size and deposit distribution. The tracked trolley carries the test plants at a controlled speed through a curtain of spray of known characteristics. Droplet size is determined by the position of the slot aperture in relation to the rotary disc atomizer. The number of droplets deposited is governed by the number of passes of the plants through the spray. Pest organisms which can be used for bioassay include powdery mildews, mites and moth caterpillars. (Photograph, Long Ashton Research Station)

surface. The target can be mounted on a moving belt or trolley. Droplet size can be varied from about 40–400 μm and the droplet density increased by passing the target through the spray curtain more than once as necessary. An air motor can be used as an alternative to an electric motor to drive the disc to produce aerosol droplets (Uk, personal communication).

Vibrating needle

Vibrating needles have been used to produce uniform droplets (Dimmock 1950; Davis 1951; Amsden 1960; Wolf 1961; Mason, Jayaratne and Woods 1963) but careful adjustments are needed to obtain a reliable stream of droplets. More recently, Merritt and Drinkwater (1977) described a laboratory device (Fig. 5.6) for producing single droplets in uniform sizes ranging from 150 to nearly 500 μm diameter. The apparatus consists of an aluminium bar and counterweights pivoted at one end and with a standard draughtsman's compass needle fixed at the other so that its angle of inclination can be altered. The needle is moved down into a liquid reservoir consisting of a flat-based glass tube 25 m diameter × 50 mm height, by a snail cam, acting on a rectangular nylon block mounted on the arm. The cam is rotated by small crank-handle so that as the heel of the cam passes the nylon block, a small coilspring under the bar allows a rapid recoil of the bar. This causes the needle to be withdrawn rapidly from the liquid in the reservoir, and form a ligament from which a single droplet is thrown in a parabolic trajectory. The surface to be treated is held so that droplets are intercepted by their downward flightpath. The vibrating needle has been used to study the effects of form of deposit on herbicide performance (Merritt 1982a, b). Smaller droplets can be obtained with a finer needle.

Microtip nozzle

A twin-fluid microtip nozzle, similar in principle to the nozzle used in the Potter tower, has been adapted to produce as near monosized droplets as possible (Uk 1978; Scopes 1981). Coggins and Baker (1983) describe an improved version of the nozzle. Munthali (1981) produced droplets within the range 20–150 μm in diameter from a fine central stylus projecting through the orifice of a G30 needle, by adjusting a flow of CO_2 from 200 to 400 cm/min with a needle valve. A flowmeter is needed to measure the flow of gas accurately. Raising the stylus within the orifice tended to produce smaller droplets. A fluorescent tracer can be added to the spray (Munthali and Scopes 1982) so that the deposition of individual droplets is seen under ultra-violet light. The target leaf discs can be moved manually under the spray for different periods of time to achieve a range of droplet densities (Fig. 5.7). Regular cleaning of the nozzle in acetone was necessary and while not in use the nozzle was stored, submerged in acetone, in a sealed jar. The microtip nozzle has been used to investigate the effect of droplet size and

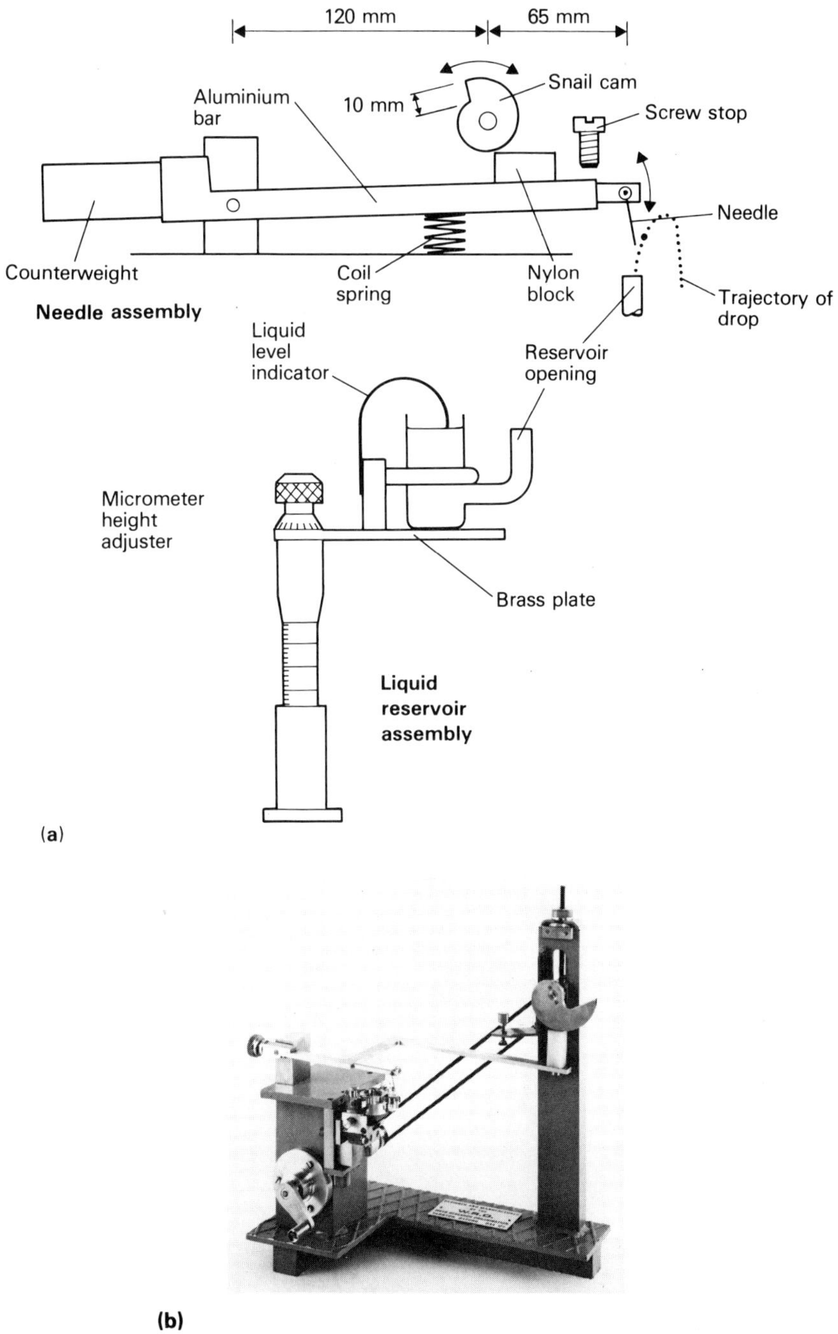

Fig. 5.6 Device for producing single droplets of uniform size
(a) Diagram to show parts. (From Merritt and Drinkwater 1977)
(b) Photograph of apparatus. (WRO photograph)

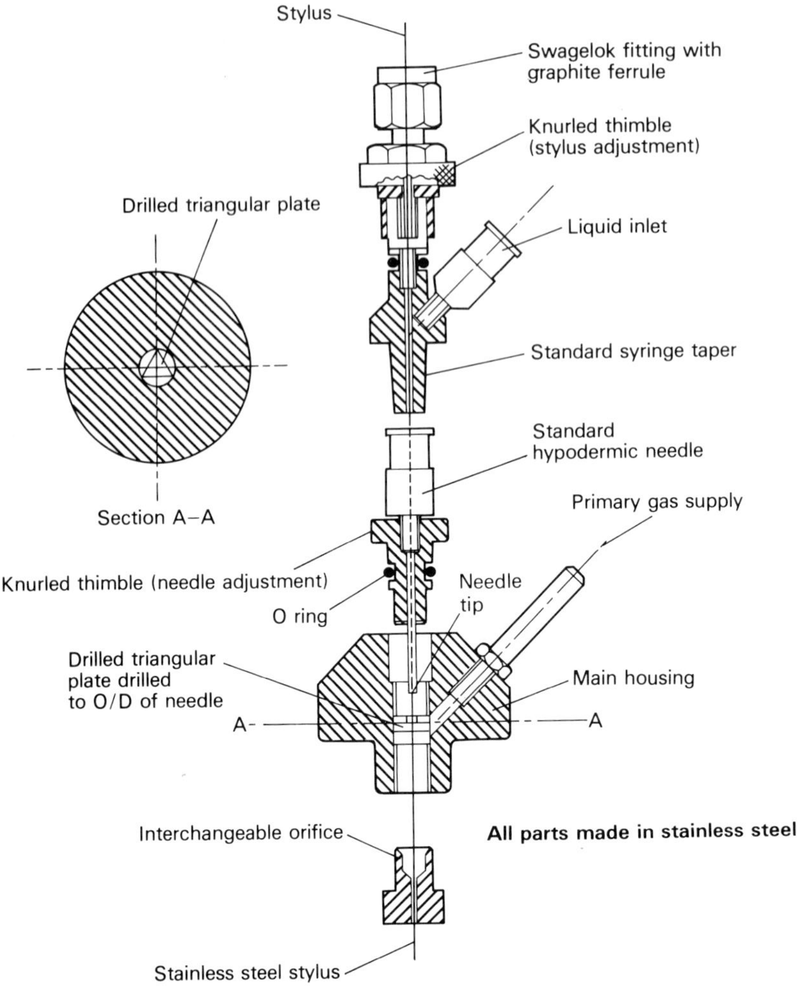

Fig. 5.7 Twinfluid nozzle to produce uniform droplets. (From Coggins and Baker 1983)

spray concentration on mortality of red spider mite *Tetranychus urticae* and whiteflies on leaf surfaces (Scopes 1981).

Spray tunnel

Aerosol droplets are usually filtered from airstreams by flying or sedentary insects, so wind tunnels have been devised through which air with droplets is drawn by a fan. When insects are placed inside cages, the mesh collects much of the spray so there is a need to devise methods of releasing flies for a brief exposure period and then recapturing them for placing in clean cages. Wooten and Sawyer (1954) describe a tunnel 30 cm in diameter with a flared opening in which the spray nozzle was mounted. Locusts were suspended in an airstream with a velocity up to 7 m/s. MacCuaig (1962) modified the tunnel with a shutter device to stop spray abruptly and used a microburette sprayer. Hadaway and Barlow (1965) used a simple plastic cylinder, 30 cm in diameter and just over 1 m in length to investigate the impaction of 20, 40 and 60 μm droplets on dead tsetse flies painted white and mounted on paper discs or cylinders to represent leaves and twigs.

Recent studies have used fluorescent tracers to examine droplet deposition on flies examined with a scanning electron microscope with cathode luminescence.

Mobile nozzle systems

Screening trials have often used a turntable (Simanton and Trammel 1965) or conveyor belt to move plants past a fixed nozzle. Many workers have used a twin-fluid (paint spray) nozzle operated at various air pressures, but as the droplet spectrum is different from nozzles used on tractor or aircraft sprayers, standard fan or cone nozzles can be used. Recent studies have involved mounting an electrodynamic nozzle (Coffee 1981) instead of a hydraulic nozzle.

Another technique is to mount the nozzle on a track so that it can be repeatedly moved at a constant velocity past targets, including potted plants. Ideally, the speed should be the same as used by field-sprayers and this can be achieved, for example, by a linear transporter consisting of a sealed tube in which a small polymer-bonded metal slug 'mole' is pushed to and fro by compressed air. A shuttle mounted on rollers is moved along the tube by a set of permanent magnets which keep in step with the 'mole' by a magnetic coupling (see p. 140). When only a relatively small area can be sprayed at a particular velocity a nozzle can be mounted on the arm of a pendulum (Amsden, personal communication). The velocity over the target can be adjusted by altering the weight on the pendulum and the distance over which it has to swing.

All the above spray equipment is usually enclosed in a suitable cabinet or a small room used specifically for spraying, from which the air can be extracted through appropriate filters to remove excess pesticide. The larger equipment usually has an automatic system for washing the inside of the cabinet and the most sophisticated units are built into an air-conditioned air duct so that air temperature, humidity and wind speed can be controlled. Separate equipment should be used for insecticides, fungicides, herbicides, etc. to avoid cross-contamination of chemicals.

Piezoelectric nozzles

Uniformly-sized droplets can be produced by pulsing spray liquid through an orifice (Frost and Yates 1981) (see p. 124). The nozzle is not suitable for the application of particulate formulations. Individual nozzles can have one or more orifices.

Calibration

With each of the techniques described above, careful calibration of the equipment is essential, together with a check on the uniformity of deposits. This can be done by incorporating a dye in the spray, washing-off the deposits and measuring the concentration of dye obtained with a spectrophotometer. Fluorimetric techniques are an alternative, or actual pesticide deposits can be measured by chromatographic or other sufficiently sensitive analysis.

The size of droplets deposited on targets can be measured by using slides coated with magnesium oxide to collect spray. Craters formed by the droplets are measured with a microscope fitted with a graticule, image-shearing system or with an image-analysing computer. Calculation of the volume and number median diameters is described by Matthews (1979). Merritt and Drinkwater (1977) collected aqueous droplets in a viscous silicone fluid and measured them with a calibrated microscope. Droplet spectra obtained from the nozzles can also be measured using lasers (Combellack and Matthews 1981; Parkin and Wyatt 1980). Although the equipment is very expensive, these systems require less labour and time and are less subjective in determining droplet size.

Aerosol sprays

Spray chamber

There is a standard specification for testing the insecticidal efficiency of aerosols produced by pressure packs against flies and certain other flying insects. The pressure pack is continuously discharged as one operator walks down the centre of the long axis of a test-room. The volume of the test-room should be about 50 m^3 and uniformly illuminated by fluorescent tubes to a light intensity of 108 m cd when measured 1 m above the floor. A temperature of 27 + 1 °C and relative humidity of 50 ± 5 per cent should be maintained and the room ventilated between tests by a fan displacing at least 10 m^3 of air per minute. Spray should not be deposited on the walls or ceiling and the floor should be covered for each test with a new layer of absorbent paper. When the pressure packs are fitted with metering devices, they are operated at several positions in the room, but in each case the nozzle must be at least 1 m from any surface. The containers are weighed before and after use, to determine the amount discharged. Approximately 500 flies are released at floor level immediately after the aerosol has been sprayed and the number knocked down counted at 2-minute intervals up to 10 minutes. The room is then ventilated and at the end of the exposure period all flies are collected and placed in clean containers with food (5 per cent sugar solution on cotton wool in a glass phial). Total mortality is recorded after 24 hr. Experimental work on which the specification is based, is given by Goodwin-Bailey, Holborn and Davies (1957).

Another widely-used test devised by Peet and Grady in 1928 uses a 6 m^3 cube and is most useful for evaluating aerosols containing pyrethrins, though it is of doubtful value with other insecticides (Busvine 1971). Other chambers have been described, e.g. by David (1946a, b).

Immersion methods

Sometimes a simple technique for comparing pesticides is needed, and this can be achieved by immersing totally the test organism in a pesticide liquid. The immersion period is kept constant and the mortality caused expressed in relation to the concentration of toxicant in the dip. An example of a dipping method for *Diparopsis* eggs is given by Matthews (1966). Eggs laid on filter paper were incubated at a constant temperature (25 °C) and on the day before hatching was expected, individual eggs attached to the paper were dipped in an emulsion or suspension. The dipped egg was then placed on clean filter paper which removed the excess liquid. The criterion for effectiveness was whether the young larva which chewed its way through the eggshell was able to hatch from the egg. This test is not suitable if the eggs split open irrespective of the amount of toxicant adhering to the eggshell. Ovicides have also been evaluated by immersing eggs cut from leaves or twigs (Steer 1938). Dipping tests for foliage-feeding larvae have not been particularly successful and topical applications are preferred. Voss (1961) and Dittrich (1962) have developed a slide-dip method for spider mites. A strip of double-sided adhesive tape is attached to a microscope slide and fifty mites, dorsal side down, are stuck to the tape in five rows of ten, by the use of a fine brush. The mites are dipped for 5 sec in serial dilutions of acaricides, then drained before being kept in a constant temperature (27 °C) cabinet at 95 per cent RH. Mites moving when prodded with a fine brush were recorded as alive at 24, 48 and 72 hr after treatment.

Aquatic larvae

The WHO test for resistance in mosquito larvae involves distributing about 300 healthy late third- or early fourth-instar larvae, in batches of 20–25 in small beakers containing 25 ml water. Stock solutions of insecticides are diluted to form dispersions with concentrations ranging from 0.004 to 2.5 p.p.m. in 500 ml beakers to which the larvae are added, and stirred. Mortality is counted after 24 hr at 25 °C. Other methods are given by Busvine (1971). A bioassay with mosquito larvae, usually *Aedes aegypti*, has been used to detect residual traces of insecticide. The residue is extracted from the sample with a volatile solvent and diluted before adding to water. Burchfield, Hilchey and Stores, (1952) used second-stage larvae.

Application to individual insects

Micropipettes

Constant volumes of pesticidal solutions can be delivered by micropipettes, constructed from narrow glass tubing with a sharply defined constriction a short distance from the tip. The tube can be filled by capillarity up to the constriction, and the liquid expelled by blowing down an aspirator tube. Hewlett and Lloyd (1959) simply used short lengths of fine capillary tube in suitable holders, but this technique is a little difficult to use. The Drummond microcap is a commercially available constant-volume pipette having a rubber bulb pierced with a small hole. The pipette is filled by capillarity and the hole closed by a finger when the bulb is squeezed to eject liquid from it (Busvine 1968). MacCuaig and Watts (1968) adapted this microcap to apply variable doses. A linear scale is fixed to the side of the capillary tube with Araldite and after filling the tube, excess liquid is removed by touching the tip on moistened cotton wool, until the required volume is indicated on the scale (Fig. 5.8). All these micropipette techniques tend to be time-consuming and for this reason cannot always be used when many insects have to be treated.

Microsyringes

A number of precision-made glass syringes are available commercially. The Agla and Hamilton syringes are suitable and are fitted with a hypodermic needle or preferably a cannula (0.3–0.4 mm diameter). The plunger is moved by the piston of a micrometer (Fig. 5.9). As repeated adjustment of a micrometer is tedious, various mechanisms have been designed to simplify regular dosing. The Arnold (1967) micrometer has a cylinder with five rings each having different numbers of depressions. The dose is regulated from 0.25 up to 5 μl, by the position of a spring-loaded ball which fits against the appropriate ring. A more expensive electrically-operated version provides doses from 0.1 to 1.0 μl (Arnold 1965). A switch activates a cam, one revolution of which pulls the toothed wheel the appropriate distance by a ratchet. A rapid rewind mechanism is used to return the micrometer screw when it becomes fully extended. Hewlett (1954, 1962) combined a micrometer with a twin-fluid nozzle so that the flow of air detaches a spray droplet and impacts it on an insect. Needham and Devonshire (1973) adapted a microsyringe to apply small drops of aphids (see microtip nozzle p. 96).

Stomach poisons

Mortality with certain insecticides, including *Bacillus thuringiensis* and the baculoviruses depends on ingesting spray deposits applied to leaf surfaces or equivalent food. The proportion of insects killed will be affected by the amount of food ingested and whether the insecticide has any contact activity

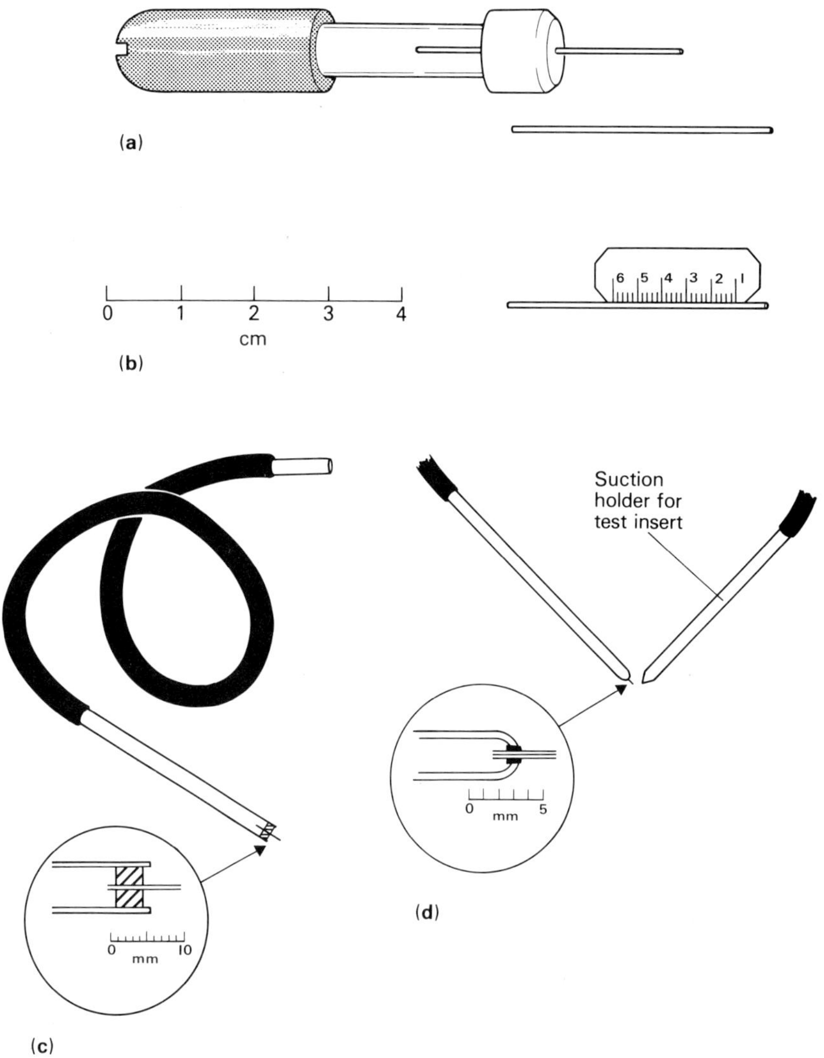

Fig. 5.8 Micropipettes for dosing small arthropods. (From Busvine 1971)

(a) Drummond 'Microcap' device. The black rubber bulb has a hole at the end, which is closed by the finger, when expelling liquid from the capillary

(b) Capillary with scale for precise dosing. (After MacCuaig and Watts 1968)

(c) Constant volume micropipette (bore 0.1 mm diameter, droplets volume 0.35 μl) suitable for insects the size of houseflies

(d) Similar micropipette for mites or very small insects (bore 0.01 mm; droplets volume 0.00045 μl)

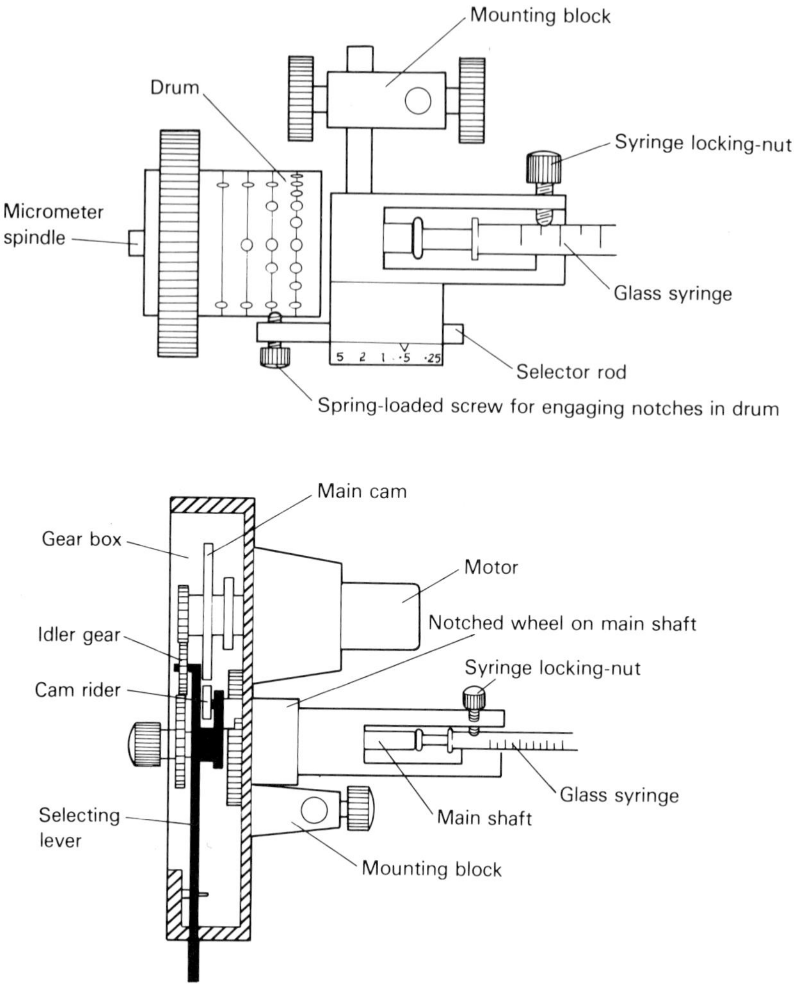

Fig. 5.9 Microsyringe applicators to apply small volumes repeatedly
(a) Hand-operated multiple-notched drum device. (After Arnold 1967)
(b) Electrically operated device. (After Arnold 1965)

as well as stomach poisoning. Ideally, the dose taken by individual insects should be known and this can be done in some cases by feeding a known number of small pieces of leaf each with a known dose. Alternatively, leaf discs with an even deposit are fed to the insects and the insecticide consumed deduced from the proportion of the disc left behind. Stevenson (1969) developed a method for measuring the toxicity of pesticides to honeybees. Groups of ten bees are placed in cylindrical cages (approx. 11×4 μcm) and while anaesthetized a 1.0 μl drop of the test material in acetone is placed on each ventral thorax using a micropipette. Oral toxicity tests used appropriate concentrations of the test material obtained by dilution in a mixture containing 5 per cent acetone and 20 per cent sucrose (Stevenson, Needham and Waller 1978).

Systemic insecticides

Sucking pests such as aphids and red spider mites can be controlled with insecticides which are absorbed by the plant and are distributed systemically up through the xylem to the leaves. Poisoned aphids falling from leaves can be collected to determine the effect of feeding them to predators. In toxicity trials the insecticide can be applied to the soil in a pot, although the amount absorbed by the roots will be affected by the type of soil used, clay soils being very sorptive and reduce systemic effects more than sand. This problem can be overcome by cutting the taproot and immersing it in a tube containing the pesticide solution (Fig. 5.10a). In practice, since much of a systemic insecticide is wasted in the soil, it can be applied as a foliar spray or, when very young seedlings need protection, seed treatment is possible. This treatment method works satisfactorily with some crops, especially those with large seeds such as beans, but may not always be suitable. Some compounds are phytotoxic to very young seedlings at the dose levels necessary to maintain a toxic effect in foliage at a later date. To overcome this band treatments under, above or beside the seed drill may be applied

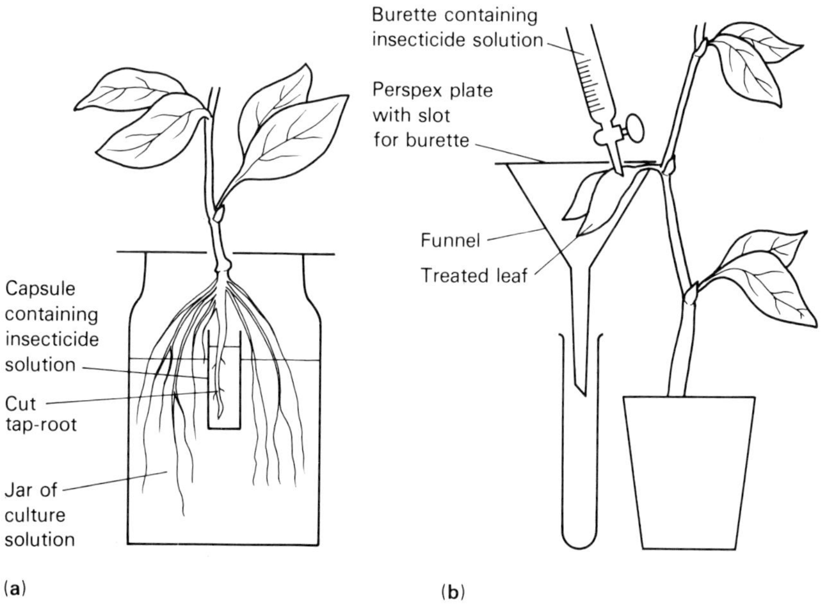

Fig. 5.10 Evaluation of systemic insecticides in plants
(a) Via cut tap-root
(b) Via the leaves
(From Busvine 1971)

(see application, Ch. 6). Insects can be confined in a clip-on cage (Fig. 5.2) to assess individual leaf toxicity, alternatively, treated leaf discs may be removed and placed on blotting paper resting on a sponge or wick from a dish of water or nutrient solution. The later technique or, when space permits, the use of whole leaves with their petioles dipped in water (Gratwick 1963) is better when translocation across foliage is to be measured since wetting the deposits with sponge or filter paper alters the characteristics of the deposit and generally enhances uptake (Galley 1968). An alternative is to cover a part of a leaf while a spray is applied; then after removing the cover, the insects are caged on the covered part of the leaf. *In vitro* tests are possible, e.g. when aphids were allowed to feed on artificial nutrient solutions having different amounts of toxicant and held in-between artificial membranes (Mittler and Dadd 1964). However, the technique should be used with caution as the test gives no information on the uptake of an insecticide from an actual leaf as it is known that plant aphids tend to take larger volumes, from plants.

Volumes imbibed from diets and, hence, the amounts of toxicant, may be determined by including a radioactive-labelled tracer or, better, by using radioactive-labelled insecticide (Galley and Gohain 1981; Ho and Galley 1982).

Statistical analysis

The biological assays described here are based on quantal response, and the proportions of each batch which react in a particular way are recorded. If the responses are plotted in relation to the different dosages, a sigmoid curve is produced (Fig. 5.11) from which it is impractical to determine the dosage which just gives 100 per cent response. It is also not easy to determine the

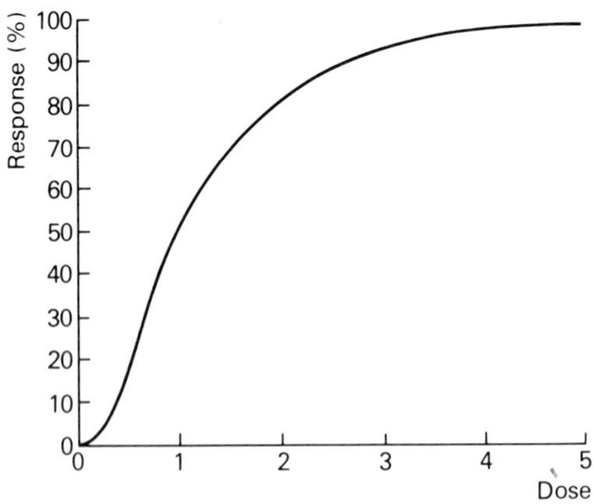

Fig. 5.11 A typical assymetric sigmoid curve relating percentage response to dose. (From Hewlett and Plackett 1979)

point of inflection at the 50 per cent level unless very many data points have been produced; even then mathematical treatment of the data to fit a curve of this type is complex. However, if the data are further transformed such that the graph approximates to a straight line, mathematical analysis is possible, and for a quick assessment of results, a line can also be fitted by eye.

The transform applied to the mortality data is based on normal standard deviates to the normal population response to log-dose (Fig. 5.12). Further details of the statistical implications are given by Hewlett and Plackett (1979). In practice, a table is used to convert the percentage mortality to a probit which, by straightening the sigmoid relationship, allows a straight line to be drawn through the points (Fig. 5.13). The equation to the line can be derived from this but for statistical reasons a linear regression will not give quite the correct result. This must be done by a probit analysis. Finney (1971) gives a full account of the use of probit analysis developed originally by Bliss (1935) with mortalities expressed as probits and dosages (or doses) converted to a log scale. There is another transformation of mortalities (logits) based on a logistic tolerance distribution rather than a normal distribution as used with probits. It is very similar to probits and has the advantage that analysis of data as logits is more easily done with modern programmable calculators (Hewlett and Plackett 1979), than a probit analysis for which most workers rely on computer programs such as Davis (1971). For those without a computer, a method of calculating the LD_{50} is given in Appendix 1.

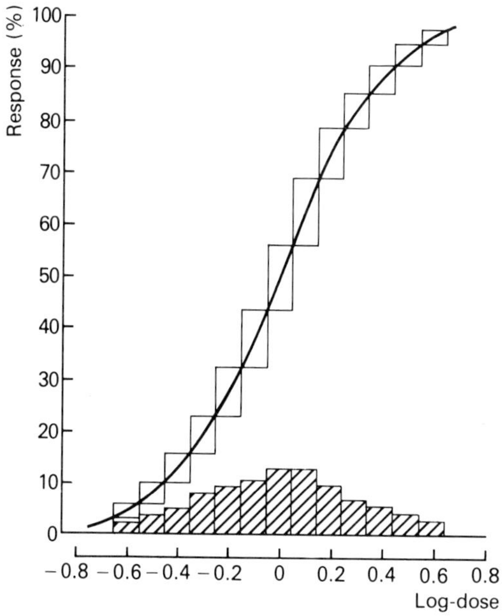

Fig. 5.12 A step diagram showing how the sigmoid curve relating percentage response to log-dose is a cumulative normal curve. (From Hewlett and Plackett 1979)

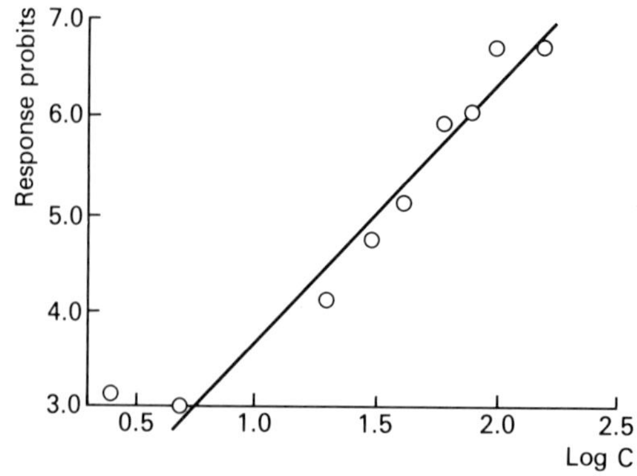

Fig. 5.13 A linear regression between probit response and log-dose. (From Hewlett and Plackett 1979)

Quite often during an experiment a proportion of test individuals may die from natural causes or for other reasons not associated with the experiment. This mortality should be as low as possible and must be measured by having untreated controls. Mortality in the other treatments is then corrected for the control mortality using a formula commonly known as Abbott's formula (Abbott 1925):

$$P_t = \frac{P_o - P_c}{100 \times P_c} \times 100$$

where P_t = corrected mortality
P_o = observed mortality
and P_c = untreated control mortality, all expressed as percentages

Where control mortality is high, the results are less significant and experiments should be redesigned if control mortalities exceed 10 per cent. Finney (1949) provides a table of weighting coefficients to adjust the data, and Finney (1971) gives the procedure for analysing data (see Appendix 1).

Having drawn or calculated the regression, the dosage required to produce the median (50 per cent) response level of a population is calculated. The dosage needed to achieve the median response in terms of mortality is referred to as the LD_{50} (i.e. lethal dosage for 50 per cent of the population) ED_{50} (effective dosage) or MLD (median lethal dose). Similar parameters include the LC_{50} for lethal concentration, LT_{50} for lethal exposure times, KD_{50} for knockdown dosage. The toxicity or potency of an insecticide is lower as the LD_{50} increases.

Other dose levels can be derived similarly and for field work LD_{95} or LD_{99} are commonly used. However, because the slope of the line will have some variance, the values near the middle will be assessed with the greatest precision (Fig. 5.14) and it is usual in comparing the effectiveness of compounds to compare the MLDs. One further proviso, however, is that for

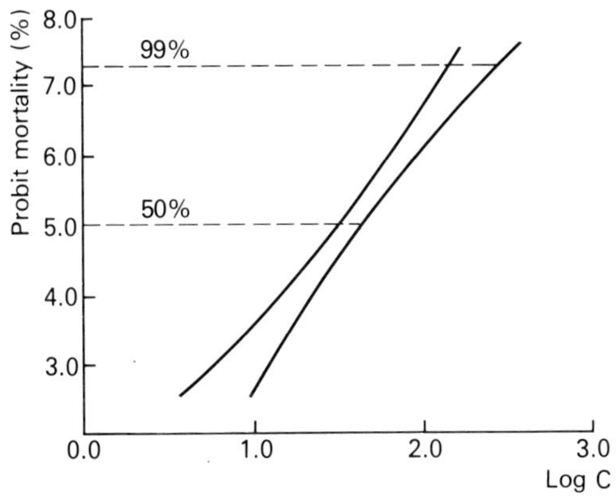

Fig. 5.14 The 95 per cent confidence limits for a regression as shown in Fig. 5.13. The variance of the slope is shown to illustrate the greater error in estimating doses above or below the median lethal dose

relative potency estimates of two or more compounds the lines must be parallel or very nearly so.

Data do not always fit the sigmoid curve illustrated in Fig. 5.11, and care must be taken in the interpretation of the result. When a population has been exposed to pesticides, the proportion of resistant and susceptible individuals may change and bioassays over a period of time in that population could give the results shown in Figs 5.15 and 5.16. Hewlett and Plackett (1979) discuss the detection of resistance using a dose which gives a very high kill (e.g. 99 per cent) of susceptible populations which is then selected to provide baseline data. This is regarded as a discriminating dose which is applied to subsequent samples collected from the field. The probability that

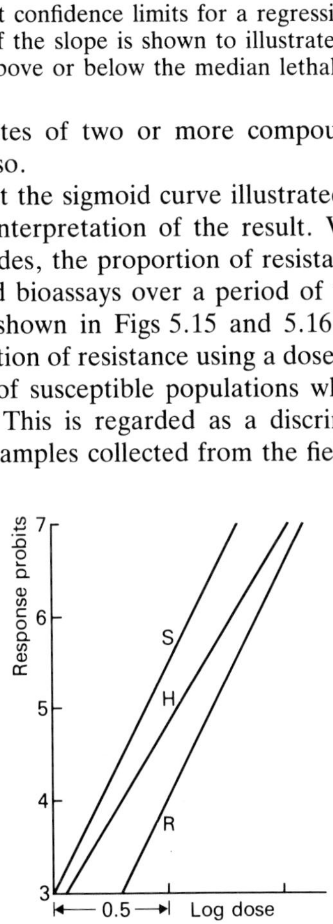

Fig. 5.15 The probit–log-dose curve, where H is from a heterogeneous population consisting of a mixture of equal numbers of a susceptible, S, and resistant, R, individuals giving straight probit–log-dose lines separated by 1.5 probits. (From Hewlett and Plackett 1979)

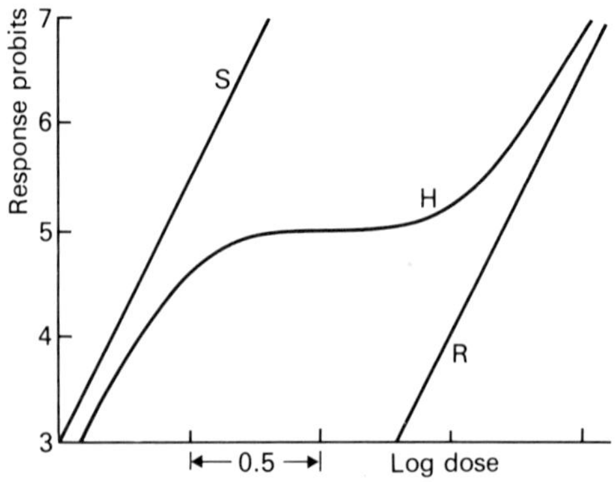

Fig. 5.16 The probit–log-dose curve H, from a heterogeneous population consisting of a mixture of equal numbers of S and R individuals giving probit–log-dose lines separated by 6.5 probits. (From Hewlett and Plackett 1979)

resistance is developing in the population is detected when there are a few survivors in, say, three successive samples. Further tests are then needed with a series of different doses to ascertain the level of resistance. Early warning of resistance is essential so that an alternative method of control or a change to another pesticide with a different mode of action can be made before the currently recommended pesticide becomes ineffective. A higher dosage of a pesticide should not be used as this is likely to increase very rapidly the selection of a more resistant population. The *discriminating dose technique* can also be used when insufficient samples are available to calculate a probit line and LD_{50}. Valuable data can be obtained in these circumstances by using discriminating doses to assess the effectiveness of candidate pesticides, and establish whether a particular chemical should be tried in the field.

Methods of detecting resistance in a pest population are important, since a failure of a pesticide may be due to a number of other causes, such as bad timing in relation to the pest infestation, incorrect dosage applied, deterioration of pecticide, especially if stored for too long in a hot climate, incorrect application resulting in poor coverage, unsuitable weather following application or poor water quality affecting the spray. Busvine (1980) has listed test methods applicable to twenty important agricultural and stored product pests, including the rice stem borer (*Chilo suppressalis*), rice leafhopper (*Nephotettex cincteceps*), cocoa mirids, Colorado potato beetle, codling moth, aphids and whiteflies. Test methods for detecting and measuring resistance of major pest species of stored cereals to the fumigants methyl bromide and phosphine are also given while Champ and Dyte (1976) reported on the susceptibility of stored grain pests. The level of resistance has to be related on the susceptibility of a 'normal' strain of the pests. It may be possible to use a strain which has been cultured for a long period in the laboratory, but sometimes a reference strain has to be collected from the field in areas where little or no pesticide application has occurred. If low-level resistance already occurs, tests can only detect further changes in susceptibility. Many of the tests involved topical application using a micro-pipette, but other methods described in this chapter are appropriate for detecting resistance. A programme for evaluating insecticides used in vector control is given by WHO (1977).

The quantal response may be measured as time; thus Matthews (1966) measured the time that first instar *Diparopsis castanea* larvae wandered over treated plants from hatching until death or the time when the larva successfully penetrated a bud or boll. This criterion was selected because larvae are exposed to a contact insecticide for such a short period before they are protected within the bracts or they penetrate a bud or boll. An insecticide which acts quickly is considered to be more likely to be effective than one which takes longer to kill, as the plants often shed buds with only partial damage. The technique was extended to assess the deterioration in effectiveness on surfaces exposed to sunlight and rain (Table 5.2). Plant growth over a seven-day period provided clean foliage, so during periods of rapid growth and prolonged insect attack, it was of greater importance to apply sprays at weekly intervals than to apply a larger dosage or a more persistent formulation.

Table 5.2 Increase in time first instar red bollworm (*Diparopsis castanea*) larvae can walk over leaves treated with 0.5 per cent carbaryl spray before death occurs; plants are exposed to sunlight or rain. (Adapted from Matthews 1966)

Treatment	**Percent increase after**	
	1 week	**2 weeks**
	%	%
Not exposed to sunlight	10	60
Exposed to sunlight	70	120
	Simulated rainfall	
	0.13 cm	2.54 cm
	%	%
Exposed to rain	31	146

The extension of laboratory assays to single-plant tests provides a more rational choice of the dosage to use in field trials. In the studies on the red bollworm *Diparopsis* on cotton in Central Africa, the egg-dipping technique showed that carbaryl was very effective. The LC_{50} value was 0.0005 a.i. per cent compared with DDT at 0.179 per cent. When sprayed on single plants an effective deposit was achieved with a 0.1 per cent concentration applied at the equivalent of 200 litres/ha on mature cotton plants. As sprays are needed during periods of heavy rain, this minimal concentration was increased by a factor of 5 (i.e. to 0.5 per cent a.i.) in the recommendations to growers, so that even on large plants, the total dosage did not exceed 1 kg a.i./ha and for much of the season was less. The effectiveness of this recommendation did depend on good distribution of pesticides where the first-instar larvae moved over plant surfaces, emphasizing the need for an appropriate application method discussed in the next chapter. Changes in the field application techniques and the introduction of very small charged droplets (Coffee 1981b) now require a fresh examination of dosage levels. Initial screening trials are needed with discrete droplets rather than applying dilute sprays at the equivalent of over 200 litres/ha.

Lower volumes provide the opportunity to use higher concentrations of pesticide, although less total chemical may be needed per hectare. Some of the techniques described earlier now permit more detailed studies of droplet size, concentration and droplet density interactions. Munthali (1981) showed that when dicofol was applied to leaf discs with red spider mite *Tetranychus urticae*, the lowest LD_{50} value was obtained when a 1 per cent spray was applied with 20 μm droplets.

Merritt (1980) has made similar studies of herbicide application, but with droplets larger than 100 μm. With herbicides, too high a concentration even of the surfactant in the formulation may cause localized phytotoxicity and reduce the uptake and translocation needed to control weeds.

Further studies are therefore required to establish dosage levels appropriate for effective control while minimizing the risk of unnecessary residues. Where lower dosages than currently recommended at present are possible, timing of application is more critical and this demands appropriate application equipment that can cover the area requiring treatment rapidly and accurately when needed.

6

How is it applied?

The dosage applied to pests in the field is very much greater than that required in laboratory bioassays, since only a small fraction of the pesticide reaches its intended target. Graham-Bryce (1977) pointed out that when an insect pest in a crop canopy had to be controlled, as little as 0.02 per cent of the applied spray was effective. When foliage is the target perhaps up to 30 per cent of a chemical is retained and absorbed into the leaves. Herrington, Mapother and Stringer (1981) reported 9–22 per cent of a spray was retained on bush trees at the late dormant stage and then increased to 22–37 per cent when trees were in full foliage. Sixty per cent or more of the spray may fail to impact or be retained on foliage and reaches the soil surface at the time of application. Rain shortly afterwards may wash-off spray deposits and add to the contamination of the soil.

As transfer of pesticide to the pests is such an inefficient process, there is a need to use improved systems of application (Matthews 1981). The use of seed treatment to localize small amounts of pesticide in order to protect young seedlings can be very important in some circumstances, though even this method of application can have deleterious side-effects. Sowing cereals treated with persistent organochlorine insecticides in the spring was found to be the main source of pesticide residue accumulation in birds, leading eventually to large-scale mortality of raptors. A ban on sowing treated seed in the spring when the birds' food supplies are scarce alleviated the problem immediately. Phytotoxicity problems, insufficient uptake or the limit to what can be put on a seed, especially small seeds, e.g. tobacco, further restrict this method of application.

Granule application is particularly suitable for the more toxic insecticides which would be too hazardous to spray, and where localized treatment, e.g. in the 'funnel' of maize plants, is effective. Granular herbicides have been used in certain crop situations such as lowland rice where the granules can be broadcast by hand, but the need for specialized applicators to meter and distribute irregular-shaped particles has limited their more wide-scale use. Granules sediment rapidly so drift is avoided, but retention on foliage is generally poor. Uptake of pesticides through the roots depends very much on soil conditions.

For most crops, a granular insecticide applied at sowing is only effective during the first few weeks of seedling development. Prophylactic granule treatments have been mostly on crops such as sugar beet, when control of aphids on very small plants is necessary to restrict transmission of viruses. Most pesticides used in this way are also effective against nematodes. Granule and seed treatments are no more efficient than well-applied sprays, but are important in pest management programmes as a way of restricting treatment to a small proportion of the environment.

Most pesticides have to be sprayed on to plant or other surfaces. The aim is usually to distribute the active chemical as evenly as possible so it comes into direct contact with the pest. In many cases control is achieved if deposits are sufficiently close and can be ingested or absorbed by the pest. The conventional method of spraying, first developed over 100 years ago, involves dilution of a concentrated formulation of pesticide with water and the pumping of the liquid through a small orifice or nozzle directed towards the target. Hydraulic nozzles typically produce a sheet of liquid which disintegrates into droplets varying considerably in size (Fraser 1958) (Fig. 6.1) (Table 6.1). The size of droplets is measured in micrometers (μm)

Fig. 6.1 Break-up of sheet of liquid from a fan nozzle 65 deg. angle 276 kPa 0.37 litres/min. (Photograph, National Institute of Agricultural Engineering)

Table 6.1 Droplet sizes from hydraulic spray nozzle. 11003 at 3 bar measured with Malvern Particle Size Analyser

Droplet size range (μm)	Volume of spray in size range	
< 16	0.75	
16–20	0.35	
20–25	0.57	
25–31	0.76	
31–40	1.22	17.9
40–51	1.81	
51–65	2.69	
65–83	3.94	
83–106	5.78	
106–138	8.43	
138–178	11.00	
178–232	14.71	70.9
232–310	18.05	
310–440	18.74	
440–720	10.55	11.2
> 720	0.65	

and is usually expressed as the volume median diameter (VMD), and number median diameter (NMD). When a representative sample of spray is divided into two equal halves by volume, then half the volume is contained in droplets larger than the VMD and the droplets in the other half are smaller. When the droplets in the sample are divided equally by number without reference to their volume, the diameter of the droplet in each half are smaller or larger than the NMD. The ratio of these two parameters indicates the spread in droplet sizes in a spray; the larger the ratio the wider is the droplet spectrum whereas a ratio of one indicates uniform sized droplets. The VMD is increased when hydraulic nozzles are operated at lower pressures, but very small droplets continue to be produced. The smallest droplets fall slowly, are easily carried in the wind and can contaminate areas outside the area being treated. The volume of spray which 'drifts' is usually quite small. The problem of drift is accentuated, when the relative humidity is low, by evaporation of the water diluent so droplets become smaller, and remain airborne longer. Even minute amounts of some pesticides can have very significant effects; thus 2,4-D drift can devastate a susceptible crop such as vines or tomatoes, even when situated several kilometres away from the area that has been sprayed. Such 'drift' damage can be caused also by volatilization of spray deposits and this may occur hours after spraying has been completed, if temperatures are sufficiently high.

Most wastage of pesticide is undoubtedly due to the largest droplets which have a high terminal velocity (> 1 m/s) and fall rapidly. Even if large droplets impact on foliage, they are liable to bounce off, unless the amount of surfactant in the formulation increases retention (Brunskill 1956; Merritt

Table 6.2 Relation between toxicant content and droplet size (Adapted from Graham-Bryce 1981)

Droplet diameter (μm)	Number of droplets per litre	Amount of pesticide per droplet* ng
20	2.38×10^{11}	0.08
50	1.5×10^{10}	1.30
100	1.9×10^{9}	10.40
300	7.1×10^{7}	280.00

(Activity of deltamethrin from Elliott, Janes and Potter 1978)

	LD_{50} *ng per insect*
Anopheles stephensi	0.04
Phaedon cochleariae	0.19
Plutella xylostella	1.13

* Assuming an application rate of 20 g active ingredient in 1 litre/ha.

1980). A 400 μm droplet will contain 1000 times the dose in a 40 μm droplet, so if this is not retained on the target, it is hardly surprising that there is considerable wastage. When the LD_{50} dosage of an insecticide is less than 10 ng/insect (Table 6.2), this can be transferred to the target in droplets which are smaller than 100 μm diameter. The relative sizes of a spray droplet and the end of an insect leg on a cabbage leaf is shown diagrammatically in Fig. 6.2. Himel (1969a, b) and others, by using fluorescent particles in the spray, have demonstrated that most of the droplets collected by insects are smaller than 50 μm, but fear of 'drift' contaminating non-target areas downwind has delayed acceptance of more efficient application methods.

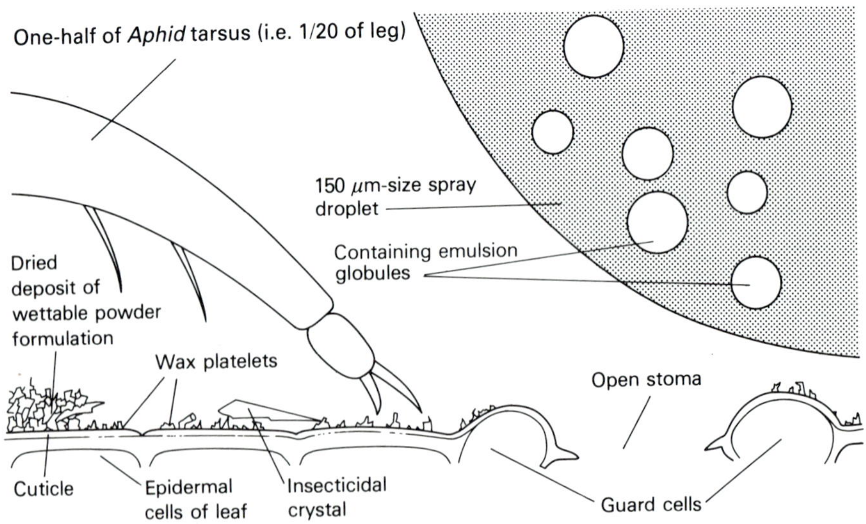

Fig. 6.2 Diagrammatic representative of some relative sizes on a cabbage leaf. (From Hartley and Graham-Bryce 1980)

Sampling sprays to show the full range of droplet sizes is very difficult. Most investigators of spray coverage have used flat paper targets, but these collect the larger droplets which sediment, and fail to show the smallest droplets carried horizontally by air movement. When an airstream meets an object, the flow is divided to pass around it. Small droplets are easily deflected and will be carried around large obstacles, but are preferentially collected on small objects which hardly change the direction of airflow. Thus the hairs on insects and narrow pine needles, for example, are efficient collectors of small aerosol droplets (< 30 (μm), which are filtered from a spray cloud. The impaction efficiency on a surface will therefore increase with 1, decrease in its size; 2, increase in the velocity of the droplet; 3, increase in the size of the droplet (Fig. 6.3). Turbulence, which mixes pockets of air containing spray droplets with air inside the crop canopy, will improve overall coverage with droplets impacted on windward surfaces in contrast to sedimentation on exposed horizontal surfaces under still

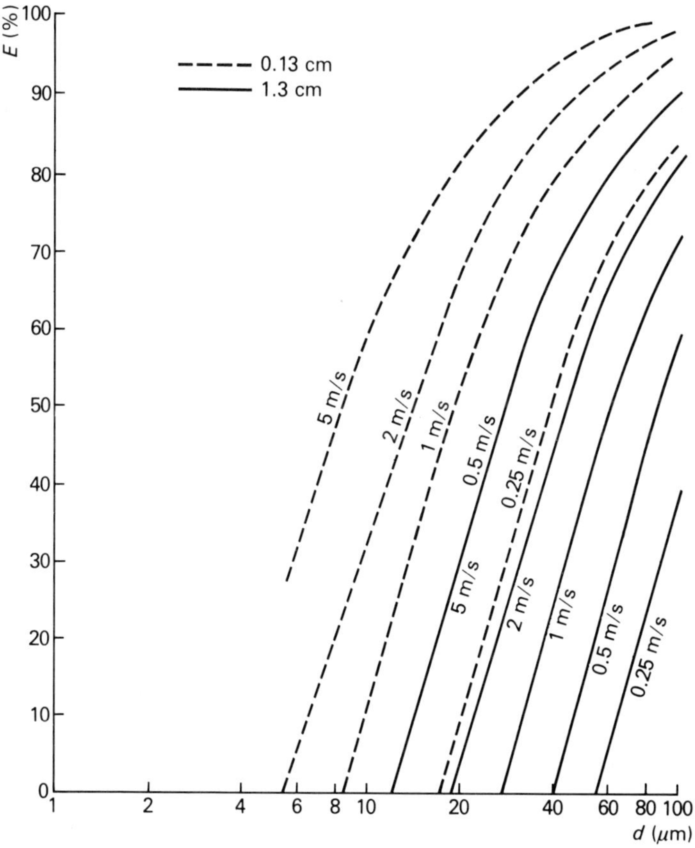

Fig. 6.3 Theoretical impaction efficiency for different droplet sizes on ribbon collectors, 0.13 cm (— — —) and 1.3 cm (— — —) wide in airstreams ranging from 0.25 to 5 m/s. (Based on May and Clifford 1967)

conditions. Sprayers with fans which can move a sufficiently large volume of air achieve better coverage on orchard tree crops.

Lawson and Uk (1978, 1979) studied the influence of wind turbulence on the dispersal of aerial sprays in relation to a wheat and fallow (ploughed) field. The large area of plant surfaces in a mature wheat crop was about 70 per cent more efficient in filtering out droplets of 50 μm than the fallow field (Table 6.3). Over a forest with trees of differing height, friction due to the crop is greater, so increasing turbulence and deposition of spray within the canopy. Increased deposition on the pine needles was obtained when fenitrothion was applied in 1 litre/ha using an involatile formulation in contrast to an emulsifiable concentrate diluted with water to apply 20 litres/ha (Joyce and Beaumont 1978).

Table 6.3 Efficiencies of wheat and fallow fields for removing spray droplets of various sizes with a wind speed 2–7 m/s. (From Lawson and Uk 1978)

Droplet sizes (μm diameter)	Wheat (%)	Fallow (%)
20	75	43
50	100	60
90	100	69
150	100	84

Controlled droplet application

Conventional spraying has been effective because the relatively large volumes of spray applied (> 100 litres/ha) wetted treated surfaces. If the wastage of pesticide is to be reduced, a narrower droplet spectrum is needed to minimize losses caused by the excessively large (> 300 μm and very small (< 10 μm) droplets. Choice of mean droplet size will depend on the particular target for the pesticide (Table 6.4). Large droplets of about 250 μm continue to be recommended where downwind movement of droplets must be avoided, e.g. when herbicides are applied, whereas droplets as small as 10–20 μm are needed to remain suspended in air to control flying

Table 6.4 Optimum droplet size ranges for selected targets.

Target	Droplet sizes (μm)
Flying insects	10–50
Insects on foliage	30–50
Foliage	40–100
Soil (and avoidance of drift)	250–500

insects as small as mosquitoes. When small droplets were deposited on leaf discs infested with eggs of the red spider mite *Tetranychus urticae*, the optimum droplet size was 20 μm with 1 per cent dicofol (Munthali 1981) (Fig. 6.4). No detailed studies with herbicides have been reported with droplets smaller than 100 μm although smaller droplets have been used to apply certain herbicides such as asulam on bracken. Lake and Taylor (1974) achieved better results with barban in 150 μm droplets compared with 250 and 350 μm droplets.

When a narrower range of droplets is applied, the spraying technique is referred to as controlled droplet application (CDA). If this type of spray is to be effective, the number of droplets deposited is as equally important as the size of the droplets. Fewer droplets per unit area should be effective if the pest is mobile, or the pesticide is redistributed and spread in some way over the surface or translocated to the site of action.

When the pest is immobile or a 'contact' chemical is applied, better

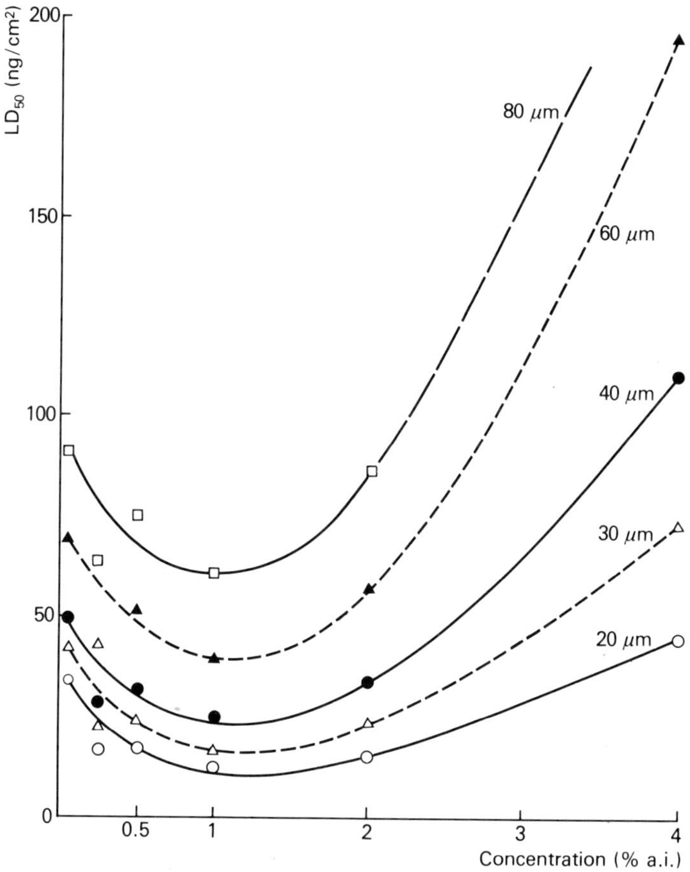

Fig. 6.4 Effect of droplet size and concentration of dicofol spray on mortality of *Tetranychus urticae* eggs. (From Munthali 1981)

coverage is needed by distributing more droplets per unit area. Halving the droplet diameter will increase the number of droplets from a given volume eightfold (Table 6.5), so smaller droplets are preferable to achieve good coverage and avoid transporting large volumes, provided the droplet size is not too small to impact. There are very definite logistic advantages in using the least volume which will give economic control, i.e. to apply ultra low volumes (ULV) of spray (Table 6.6).

Table 6.5 Change in volume rate to achieve the same droplet density

Droplet size (μm)	**Minimum volume (litre/ha) to achieve 100 droplets/cm over a flat surface**
50	0.65
100	5.24
250	81.80

Table 6.6 Volume rates for different crops (litres/ha)

	Field crops	**Trees and bushes**
High volume	> 600	> 1000
Medium volume	200–600	500–1000
Low volume	50–200	200–500
Very low volume	5–50	50–200
Ultra low volume	< 5	< 50

Collection of water can be a very difficult and time-consuming operation in many areas of the world and even when water is readily available much time is wasted while refilling the sprayer (Table 6.7). The logistic advantage of ULV spraying is important whether it involves hand-carried, knapsack, tractor or airborne equipment. This is particularly so when weather conditions suitable for applying sprays are limited. When large areas need

Table 6.7 Filling time as a percentage of spraying time based on a survey of tractor spraying on UK farms. (from ADAS 1976)

Filling time (%)	**Farms in survey (%)**
< 20	4
20–30	8
30–40	32
40–50	17
50–60	19
60–70	6
70–80	6
80	8

to be treated very rapidly to prevent a pest infestation building-up, spreading, and causing economic damage, rapid treatment is essential.

In the UK, the number of autumn or spring days suitable for spraying winter-sown cereals with herbicides is severely limited in some years, not only by the wind and rain, but also through wet soil preventing access into fields (Thompson 1982). The need to spray early to check weed growth has stimulated a demand for lightweight sprayers with low-pressure tyres to reduce soil compaction. Weight of equipment can be decreased by having small spray tanks, made possible when the volume apppication rate is greatly reduced.

Even when weather conditions suitable for spraying continue over several days, rapid application can be vitally important where sprays are only to be applied if pest populations exceed the ET level. Any delay will result in an economic loss, so large areas may have to be treated in hours. Aerial application of less than 0.5 litres/ha technical malathion has been used in emergency control operations against mosquito vectors (Lofgren *et al.* 1970). Reduced volumes have been applied with hydraulic nozzles, thus for herbicide application the '7-gallon' system (78 litres/ha) has been used successfully with 80015LP fan nozzles (Barnes 1981). Wijewardene (1982) has similarly used impact nozzles to apply very low volumes of herbicides with knapsack sprayers in Sri Lanka. These systems have the advantage that they adapt existing equipment, but filtration is critical to avoid blockage of the small nozzle orifices. Another development which eliminates the risk of wasting spray if weather conditions prevent the completion of a spray round is to keep the concentrate and diluent in separate tanks. This has been used extensively on spray trains (Amsden 1970) but so far not on farm sprayers. In Finland, a sprayer is manufactured with a dual pump metering and mixing system.

When reducing the spray volume, many operators have attempted to apply the same mass application rate of the active ingredient, thus greatly increasing the concentration of pesticide and other components in the formulation. This can have serious phytotoxic effects and prevent uptake and translocation of some pesticides. Too high a concentration of paraquat, for example, merely burns holes in the leaf. The concentration needed when reduced volumes of spray are applied has to be evaluated by laboratory or small-scale field trials, and a balance struck between the requirements for control of the pest and appropriate persistence of the spray deposits and the need to avoid phytotoxicity on the crop. Some insecticides have been applied as highly concentrated formulations, notably malathion at 95 per cent a.i., but the concentration should not normally exceed 10 per cent a.i. (Fig. 6.5).

If droplets smaller than 100 μm are applied, some control of droplet size during its trajectory is needed. The VMD of water droplets sampled at 10 cm from the nozzle at 32 °C in 20 per cent RH was 100 μm, but in the smaller sample collected when the spray reached 2 m from the nozzle the VMD had decreased to 56 μm. By adding an involatile liquid – oil – the change in droplet size was reduced (Wodagenah and Matthews 1981).

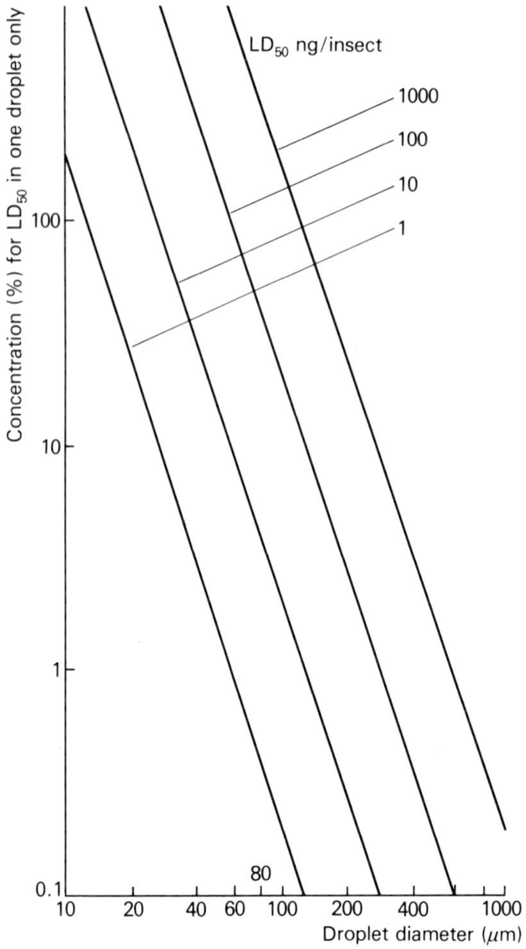

Fig. 6.5 Relation between toxicity, droplet diameter and concentration of active ingredient for one droplet to contain an LD_{50} dose. (From Johnstone 1973)

Formulations incorporating oil or glycols, such as propylene glycol are important to stabilize droplet size. When an aqueous spray is required, larger droplets are needed to offset the effects of evaporation; if small droplets are applied, each droplet should contain a proportion of involatile liquid. Thus with 12.5 per cent involatile liquid, and the remaining fraction being volatile, the size of droplets can at most be halved during their transport from the nozzle to the target.

Spinning-disc nozzles

More uniform-sized droplets can be produced using a spinning disc (Fig. 6.6). The size of the droplets is inversely proportional to the angular

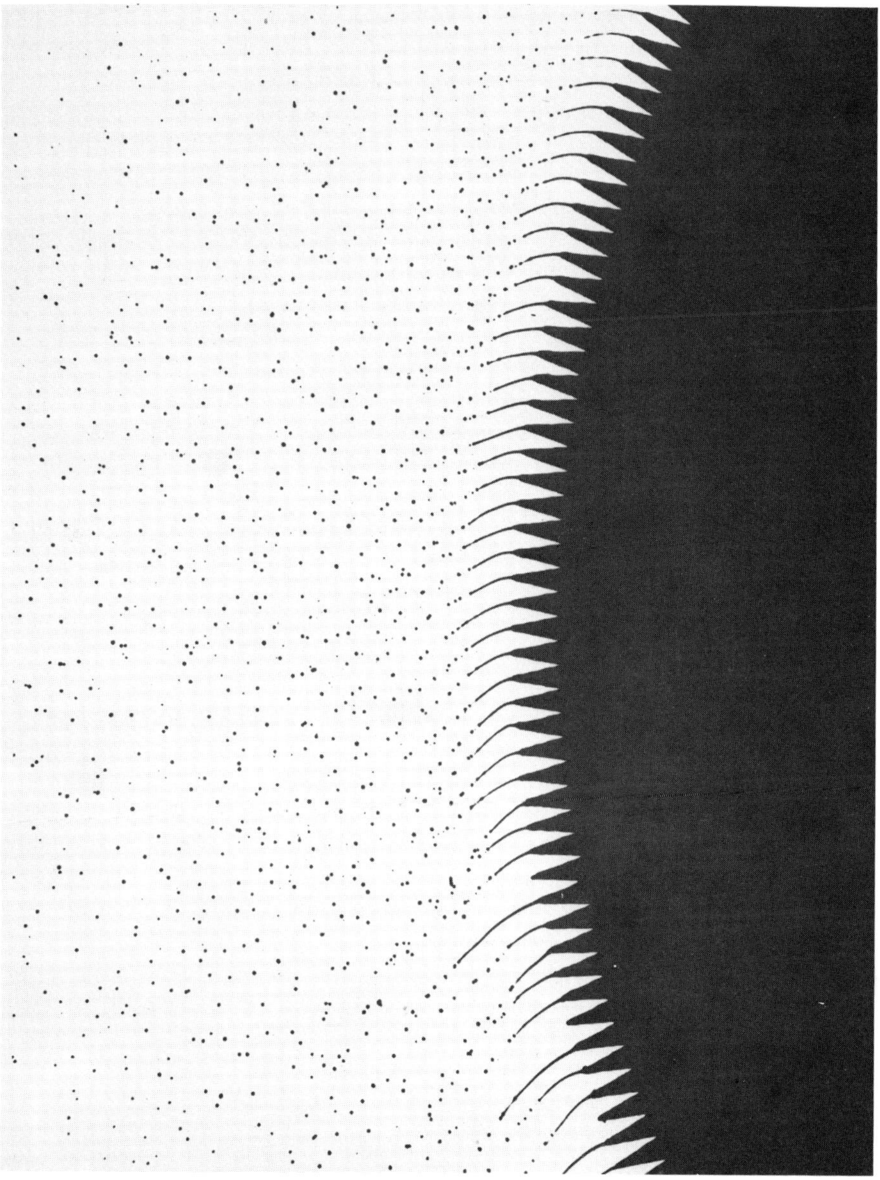

Fig. 6.6 Spinning disc nozzle showing formation of droplets from ligaments. (Photograph, Micron Sprayers)

velocity, but the actual droplet size obtained is also affected by flow rate. At very low rates, individual droplets are thrown from the edge of the disc or cup. As the flow rate is increased without changing the disc speed, some of the liquid will form ligaments which will break-up into smaller droplets and satellites. The satellite droplets will be about one-tenth the diameter of the main ones. Further increase in flow rate will eventually result in only ligaments being produced and these will become thicker and produce larger droplets until the flow rate is so great that the ligaments merge and form a sheet. The droplet spectrum is bimodal when the transition from single droplet to ligament formation occurs, so care is needed to match the flow rate and disc speed to achieve the required VMD and VMD/NMD ratio so as to have a narrow spectrum. Most commercially available discs have teeth around the circumference to minimize the effects of surface tension on droplet or ligament formation. Grooves to feed individual teeth reduce slippage of liquid over the disc surface and accommodate higher flow rates. Studies on droplet formation from spinning disc nozzles, commonly referred to as *rotary atomizers*, have been reported by Walton and Prewett (1949), Hinze and Milborn (1950), Dombrowski and Lloyd (1974), Boize and Dombrowski (1976), Heijne (1978) and Frost (1981). Spinning-disc nozzles on hand-held equipment have been described by Bals (1975a,b), Matthews (1979a, b), on tractors by Taylor, Merritt and Drinkwater (1976) and Farmery (1978), and on aircraft (Bals 1978). Rotary cages, i.e. Micronair units, have been widely used on aircraft.

Vortical nozzles

In the vortical nozzle spray droplets from a straight air-jet are shattered in a turbulent vortex created by an airstream passing a series of static air-baffles or vanes, and the aerosol droplets in the emerging airstream are generally less than 30 μm diameter. The vortical nozzle often referred to in the USA as a 'cold-fogging nozzle' has largely replaced the use of thermal fogging equipment in vector control and has been used to apply microbial insecticides (Sorensen and Falcon 1980). The nozzle is ideal when spray is directed at flying insects, but the droplet spectrum is not narrow, since a large proportion of the droplets are less than 5 μm in diameter. Twin-fluid nozzles using higher pressures of air have been used to produce aerosols or mist sprays.

Microjet nozzles

Droplets of remarkably uniform size can be produced by pulsing liquid through small orifices using a piezoelectric transducer (Bouse 1975; Yates and Akesson 1978). Droplet size is approximately double the orifice size, thus 246 μm droplets were produced with a pulse frequency of 12,000 Hz and a pressure of 41 kPa when the orifice was 127 μm diameter. As the orifice is so small, nozzle blockages are a major problem so spray solutions have to be carefully filtered. Stent, Taylor and Shaw (1981) electrostatically-charged droplets from a microjet nozzle to stop droplets from coalescing (Fig. 6.7).

Electrostatic sprays

Charged spray droplets are readily deposited on an earthed object, so total deposition on foliage can be significantly increased compared with uncharged sprays. Law (1980) reported up to seven times more spray was deposited on artifical targets (spheres) in the laboratory, but in the presence of a spike, droplets lost some of their charge and this decreased the deposition. Increased deposition on potted plants using electrostatically-charged sprays has also been reported by Arnold and Pye (1980), and in field crops including cotton and soyabeans (Arnold and Pye 1981). Morton (1982) obtained similar results on cotton.

Hydraulic sprays can be charged (Marchant and Green 1982) and charged rotary nozzles have been developed (Arnold and Pye 1980). However, the most revolutionary development has been that of the Electrodyn nozzle (Coffee 1979) (Fig. 6.8). Although special liquid formulations with a resistivity in the semiconducting range (up to 10^8 ohm/m) are essential with this nozzle, it can produce a very narrow droplet spectrum. Liquid is metered through a narrow gap in the spray head at a potential in the order of 15–35 kV. A field-intensifying or counter electrode connected to earth is placed close to the gap so that the emerging liquid is subjected to an intense divergent electrical field. The liquid forms waves, from each crest of which a ligament is produced and subsequently breaks-up into main and satellite droplets. An increase in voltage produces more, thinner ligaments and hence smaller droplets for any given flow rate. Increasing the flow rate produces thicker ligaments and larger droplets. The narrow droplet spectrum is achieved through the uniformity of ligaments, when the size of the gap is constant and flow rate is regular. The nozzle has no moving parts, uses only electrical forces and requires very little energy.

When there is a strong electrical field between the nozzle and an earthed object, charged droplets repelled from an electrostatic nozzle travel very rapidly to the crop target. Their velocity greatly exceeds their terminal velocity due to gravity. Even in a strong wind, downwind displacement of a charged spray cloud is minimal, so reducing the problem of 'drift'. This control of the droplet trajectory is more pronounced if the nozzle is held close (< 0.4 m) to the crop. However, if the spray is released too close to foliage, there is less chance for the droplets repelling each other, to spread out to form the dynamic cloud needed to distribute droplets as evenly as possible in the upper canopy. The collection of charged droplets is so efficient that penetration into a crop canopy can be poor unless an airstream moves the spray cloud through the outer foliage. This can, of course, be an advantage when pests are primarily on the more exposed outer sections of a crop, such as *Heliothis* spp. on cotton or cereal aphid *Sitobium avenae* on wheat, since the proportion of spray contaminating the soil surface is almost negligible, thus reducing the impact on natural enemies such as ants, spiders and carabids.

When penetration of a canopy is important, it may be necessary to mount an electrostatic nozzle in an airstream as on orchard sprayers, or to reduce the charge/mass ratio of the droplets. The voltage can be adjusted on most electrostatic nozzles, or partial discharge of droplets can be achieved by fixing a point, connected to earth, close to the nozzle orifice (Coffee 1980). When the charge on a spray is reduced, droplets will be carried downwind in a similar manner to an uncharged spray, though the distance over which

(a)

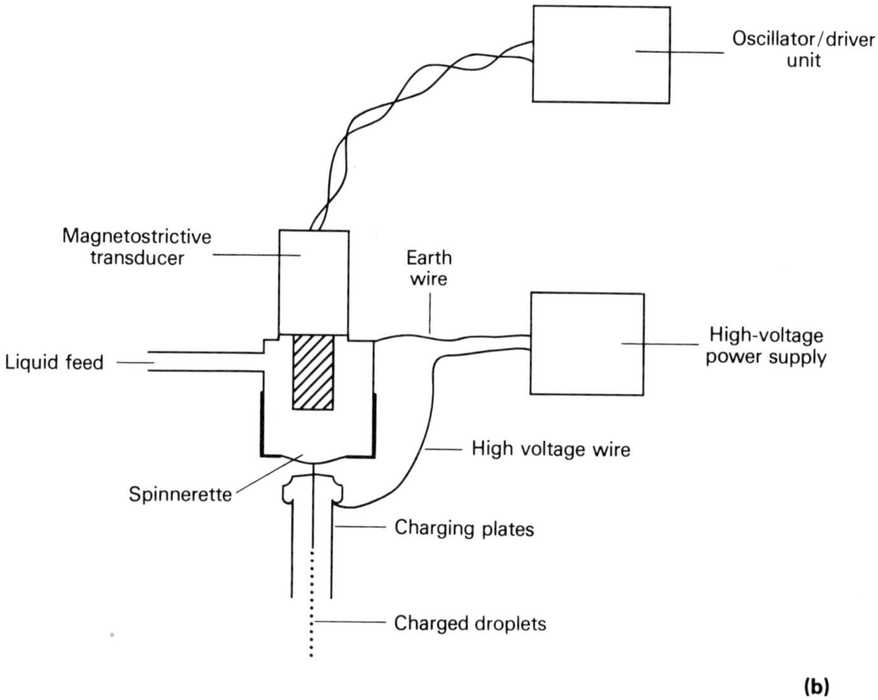

Fig. 6.7 (a) Electrostatic pulsed microjet nozzle. (Photograph, WRO)
(b) Diagram to show components of microjet nozzle. (From Stent, Taylor and Shaw 1981)

the spray cloud will drift will be more predictable if there is a narrower range of droplet sizes. The direction in which a charged spray cloud moves can also be influenced by careful positioning of a metal rod or plate, referred to as a *deflectrode* which is similarly charged to the nozzle, and which therefore will repel the charged droplets (Coffee 1981b) (Fig. 6.9). The charged surface has to be covered by a plastic sleeve to prevent anyone touching it and receiving a shock. A 'deflectrode' can be used to push the spray cloud away from the operator and achieve a wider swath. Future development of the method may include a guard to protect a crop while weeds are sprayed with herbicide, or individual rows in an inter-crop system are treated with pesticides.

An important feature of electrostatically-charged sprays is that droplets are also deposited on the undersides of leaves and on stems and petioles. Certain pests, normally difficult to kill with conventional sprays unless the pesticide is translocated within the plant, should therefore be more effectively controlled with contact pesticides. Since downwind displacement of droplets smaller than 100 μm diameter can be significantly reduced, it is also theoretically possible to maintain good coverage with much smaller volumes of spray. In contrast to the 100 litres/ha of a DDT spray needed to control the bollworm *Heliothis* spp. in the upper half of the cotton crop canopy only

2.5 litre/ha is needed when applied in a spray with approximately 70 μm droplets using a spinning disc ULV applicator. A further reduction to 0.5 litre/ha has been achieved with cypermethrin in similar-sized droplets. If the droplet size could be further decreased to 20 μm, then as little as 12 ml/ha may be needed.

The problem is whether it will be possible to achieve such low flow rates without serious nozzle blockages, and secondly whether adequate distribu-

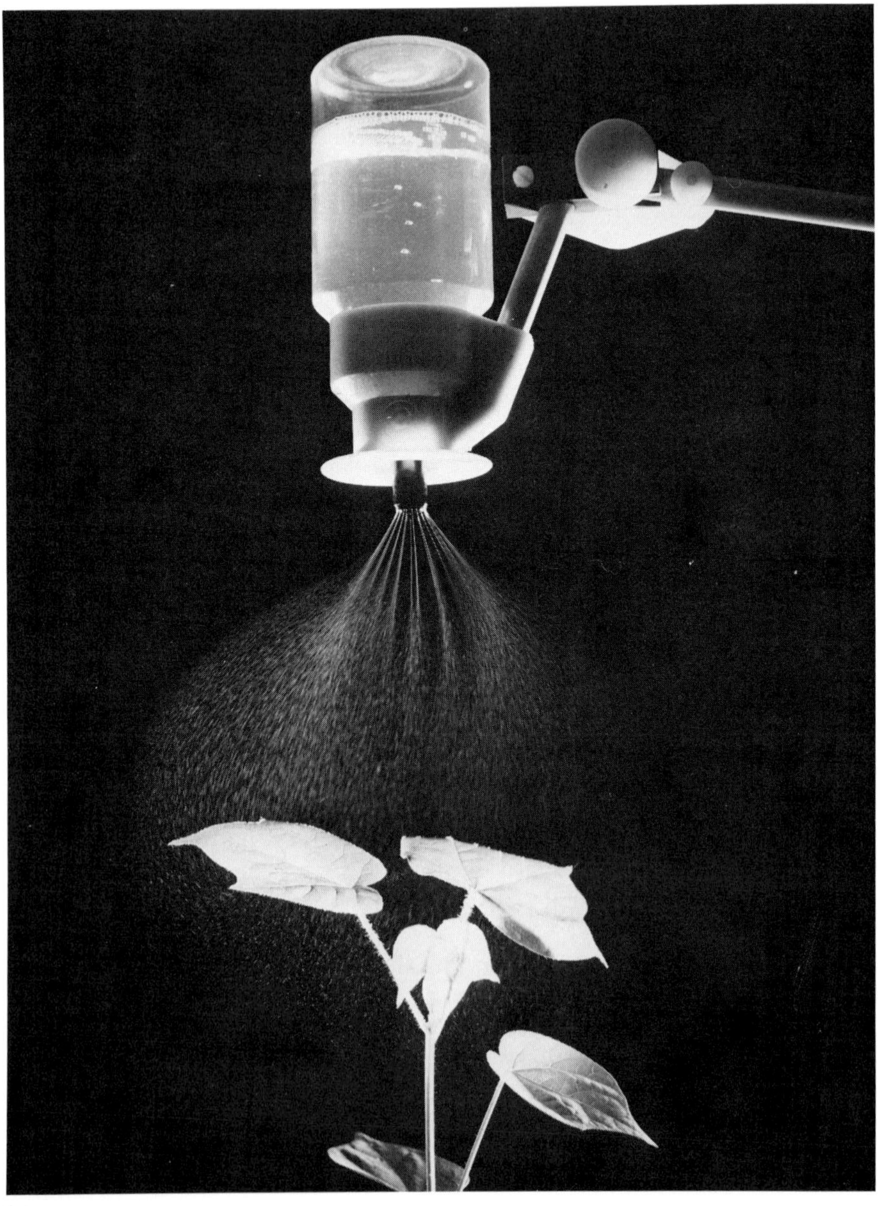

(a)

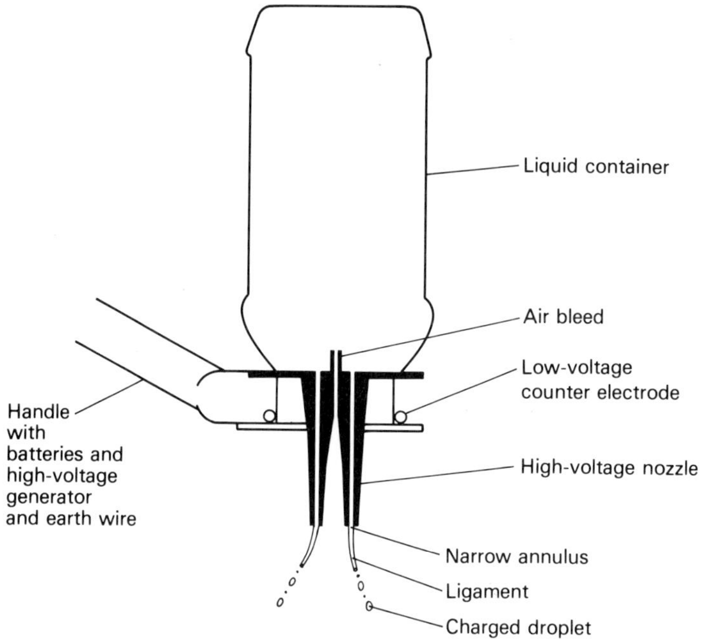

(b)

Fig. 6.8 (a) Electrodynamic sprayer (Photograph, ICI)
(b) Diagram to show components. (From Coffee 1979)

tion of the spray can be achieved. Nevertheless, there is considerable scope for investigating the use of pesticide in quantities of less than 1 litre/ha.

Improved retention of spray droplets on foliage by using electrostatically-charged sprays and eliminating the gross wastage in excessively large droplets indicates that lower dosages of some pesticides should be as effective as the currently recommended rates with hydraulic nozzles. Few results have been published, but the lower dosage of 16 g a.i./ha cypermethrin gave significantly better reduction in the number of whitefly *Bemisia* spp. 4 days after spraying than 60 g a.i./ha applied through a hydraulic nozzle (Matthews 1981). Reduced dosages of pesticide, especially insecticides and fungicides, have been effective on farms when ULV sprays have been applied with uncharged sprays from spinning discs, but the scope for reducing herbicide dosages has been generally less, particularly while applying 250 μm droplets. Notable exceptions include some translocated chemicals such as glyphosate. Further study of herbicide application with smaller charged droplets, even less than 100 μm diameter, is needed.

Operator hazards

When much smaller volumes of spray can be recommended to the farmer, it is then possible to eliminate the need for collecting water, measuring-out

(a)

(b)

Fig. 6.9 (a) Electrodynamic spray without deflectrode
(b) Electrodynamic spray with deflectrode (Photographs, ICI)

the concentrate and mixing the spray. These operations are generally the most hazardous for the farmer and have led to legislation in the USA for the provision of closed mixing systems, particularly for the most toxic pesticides. If less than 1 litre/ha of total spray is required, it can be delivered pre-mixed direct to the farmer. In the development of the Electrodyn this has been taken a stage further by the incorporation of a nozzle into the container itself, forming the so-called 'Bozzle' (Coffee 1981a and b).

This closed system can be extended to vehicle-mounted equipment and as Coffee (1981a) points out it is not difficult to envisage a sprayer of tomorrow having a high degree of electronic control combined with a microprocessor into which the farmer would key relevant information on the crop and pests. On the basis of this input the appropriate pesticide would be metered at appropriate flow rate and droplet size would be selected automatically.

Selective application

Localized treatment of pesticides is very important in pest management. Insecticide sprays may be applied to the resting sites; thus discriminative and selective residual treatments against tsetse flies were confined to the shady areas on the undersides of main branches and the lower parts of tree trunks in areas where the flies retreat in the dry season (Davis 1967). Sequential aerial application of aerosol sprays has provided large-scale control with minimal effects on non-target organisms.

Bait sprays can be used to attract a pest to the insecticide and avoid using overall coverage. Normally bait sprays are applied in large droplets. For example, dimethoate or fenthion is mixed with protein hydrolysate to attract olive flies (*Dacus oleae*). Baits may be crucial to increase ingestion of microbial insecticides, but small droplets may be needed to achieve adequate coverage where larvae feed. Sucrose was added to *Bacillus thuringiensis* against cabbage loopers *Trichoplusia ni* and a cottonseed meal bait was added to *Baculovirus heliothis* against *Heliothis zea* (Smith, Hostetter and Pinnell 1980). Smith and Bouse (1981) have pointed to the need for more research on the application of microbial pesticides.

Instead of the application of a single dose which is rapidly washed from foliage by rain or decays over a longer period, more selective application may be possible under some circumstances if the rate at which the active ingredient is released could be controlled in relation to the progress of a pest infestation. Controlled release is most important when the more volatile or rapidly degraded compounds such as pheromones are applied. Also, losses during packaging and the toxic hazards may be reduced with controlled release formulations. Cardarelli (1976) provides a comprehensive review of controlled release formulations. Micro-encapsulated pesticides have been examined with baits containing synthetic pyrethroids for the control of leaf-cutting ants which carry the poison into nests (Phillips, Etheridge and Scott 1976). However, as mentioned elsewhere there is a danger that bees will collect the microcapsules (see p. 78), and the capsules could break during storage or release the chemical when the pest is not active, so great care must be taken with formulation to achieve an appropriate release rate. The extra cost of slow release formulations has restricted more widespread use. More recently the concept of 'in flight' encapsulation has been proposed by

Cardarelli (1979) to provide a coherent envelope around droplets.

Slow-release formulations of herbicides are also being evaluated. Thus, diquat in a formulation with sodium alginate (Barrett 1978) has been applied to streams and drainage channels as 'strings' from a modified knapsack sprayer (Barrett and Murphy 1982). The technique of alginate gels allows localized control of submerged weeds (Barrett and Logan 1982). Cardarelli and Cardarelli (1982) have reviewed controlled-release pesticide formulations in aquatic environments, but there are also possibilities of using polymer films which could also reduce evaporation, especially in arid areas.

When residual deposits are applied the pest will be subjected to the pesticide over a longer period of time. This is liable to increase selection for resistant populations, so aerosol sprays aimed at adult insects may be more advantageous in a pest management programme, even though a series of sequential treatments may be needed. An example of this is the use of fogs to treat *Helopeltis* spp. bugs in cocoa in South-East Asia in order to treat large areas quickly, provided suitable conditions occur.

Herbicide application to a whole field may not be economic when only a few weeds are present, though seed from such weeds presents a long-term risk and some control is needed. When weeds such as wild oats are taller than the crop, the height differential between the weed and the crop can be exploited by using a rope-wick applicator or similar equipment which transfers pesticide to the weed as it is touched. Design and choice of materials is important to ensure adequate transfer of chemical to the weeds without it touching or dripping on to the crop. Volunteer potato plants were killed when as little as 1–2 ml of 1–4 per cent glyphosate was applied in this manner to the top 2 or 3 leaves (Lutman 1979).

Spray equipment for small-plot trials

The performance of a particular pesticide or application technique needs to be examined before it can be recommended to the farmer. Various types of equipment have been developed to apply pesticides on small plots so as to simulate as closely as possible the techniques to be used by farmers. Most equipment described in the following sections was designed to apply spray volumes over 200 litre/ha although there is a need to adapt such equipment for lower volumes and provide greater control of droplet size. Even with conventional hydraulic nozzles, increasingly greater consideration is being given to the need to have similar spray spectra in experimental plots to that on farms, so attention is being given to varying nozzle size, pressure, operating height and other factors. A change in orifice size to accommodate a change from a walking speed of 1 m/s to tractor speeds will tend to increase the VMD of the spray for a given volume application rate. Standard sprayers including knapsack and compression sprayers are frequently used for experimental plots. This may be satisfactory when the plot size is large, as the treatment can be very similar to commercial practice, especially in areas with small farms. The disadvantage is that output of spray and forward speed may not be sufficiently constant for accurate application. Provision of a pressure-regulating valve on the outlet of a compression sprayer or on a lance is needed to avoid flow rate decreasing and droplet size increasing during the application of a treatment. If a pressure regulator is not available, the pressure change can be minimized by having a large volume of compressed air and a relative small volume of liquid. Thornhill (1974) described the adaptation of a stainless steel container to provide a compression sprayer for plot-spraying. Using small readily interchangeable 6-litre containers several treatments can be applied easily, whereas most machines have too large a container.

Details of several plot-sprayers have been published in a *Directory* compiled by the International Association on Mechanization of Field Experiments (IAMFE) based in Aas, Norway, in association with the American Society of Agricultural Engineers (Quick and Buckele 1972)

Oxford Precision Sprayer

This sprayer was designed specifically for the application of herbicide to small plots (Fryer and Elliott 1954). The basic principle is similar to the compression sprayer, but the liquid and compressed air are in separate containers. The air container is carried on the operator's back while the liquid container attached to a plate is slung from the straps in front (Fig. 6.10). The 18-litre light alloy container is pressurized up to 700 kPa with compressed air at a local garage or by using a foot pump. The liquid container consists of a Pyrex glass tube secured between two metal plates, the upper one is easily removed to pour-in the spray solution, while the lower plate is fitted with a reducing valve through which air is fed during spraying. Spray liquid is forced through a hose to a lance with trigger valve and short boom for spraying plots up to 1.8 m wide. Depending on the size of the liquid container, plots up to 125 m^2 can be treated when applying herbicides at 224 litres/ha. When very small plots (0.8–1 m^2) or individual plants have to be treated during primary herbicide evaluation a smaller liquid container can be used. This microsprayer attachment incorporates regulating and trigger valves (Fryer and Elliott 1954) (Fig. 6.11). It may be necessary to use an alternative metal container where safety regulations do not permit pressurization of a glass container, although the Oxford Precision Sprayer (OPS) glass is well protected by a metal screen.

Operator fatigue has been alleviated by the addition of a counterbalance to the lance which has also been improved by allowing higher flow rates with less pressure loss (Thornton 1974). Lightweight nozzle fittings complete with diaphragm check valves are fitted so that nozzle spacing can be adjusted easily (Fig. 6.12). Ideally, individual nozzles are supplied through the same length of hosing from a central manifold.

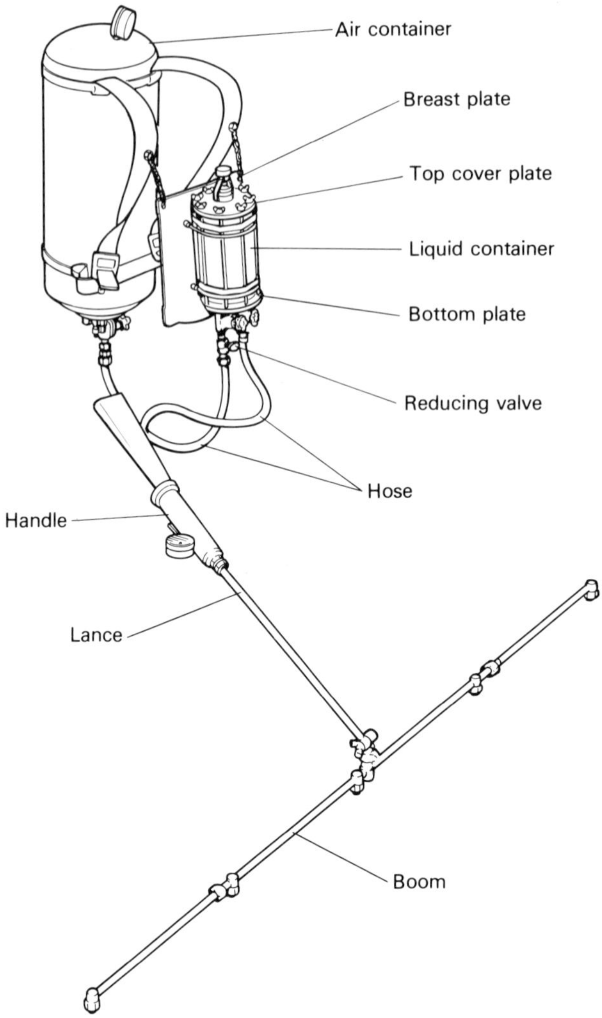

Fig. 6.10 Oxford Precision Sprayer

An adaptation to the OPS involves a shield made from three layers of nylon mesh netting to restrict wind movement (Blair, Taylor and Tottman 1975) (Fig. 6.13). Two nozzles adjustable both horizontally and vertically are mounted on a boom fixed to a four-wheeled trolley pushed by hand across the netted frame. Two operators can carry the entire sprayer between plots. Spray distribution was optimal when 8002 fan nozzles were mounted 510 mm apart and 460 mm above the sprayed surface.

Similar sprayers have been developed with quick-fit couplings so that the sprayer can be dismantled easily for cleaning and transport. On some, the spray boom is counterbalanced and offset so that the operator walks alongside rather than through the middle of the plots.

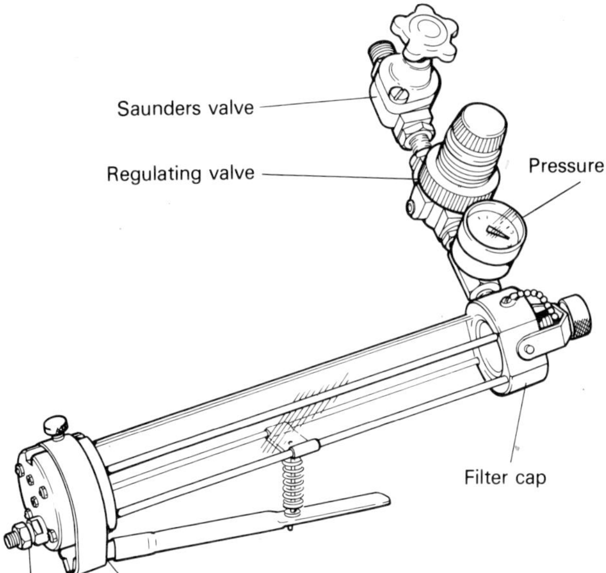

Fig. 6.11 Microsprayer attachment

Two-man plot sprayers

A portable lightweight spray boom carried by two operators, one at each end, allows wider plots up to 4 m wide to be treated (Thornton 1977) (Fig. 6.14). The boom is constructed from two lengths of square aluminium tube, supported at each end and at two other positions by vertical struts. To prevent the boom from twisting when carried, carrying-handles are mounted to the outer struts so that they are at right angles to each other, one in-line with the boom, the other pointing to the rear. One of the operators carries a backpack with a liquid CO_2 supply in a lightweight aluminium high-pressure cylinder capable of withstanding a maximum pressure of 13,800 kPa. This serves to pressurize the spray liquid carried in a second pressure cylinder. Two pressure gauges to record the pressure in the CO_2 cylinder and outlet pressure (up to 414 kPa) are mounted alongside a reducing valve connected to the shut-off valve.

A second pressure regulator is used to adjust the pressure of CO_2 before it is piped to the spray tank to force spray liquid to a trigger valve on the boom. Depending on the length of the hosing, allowance is made for losses of pressure, so the pressure of gas leaving the cylinder should be regulated to provide approximately 140 kPa more and the final adjustment of 14 kPa higher than the pressure required at the nozzles.

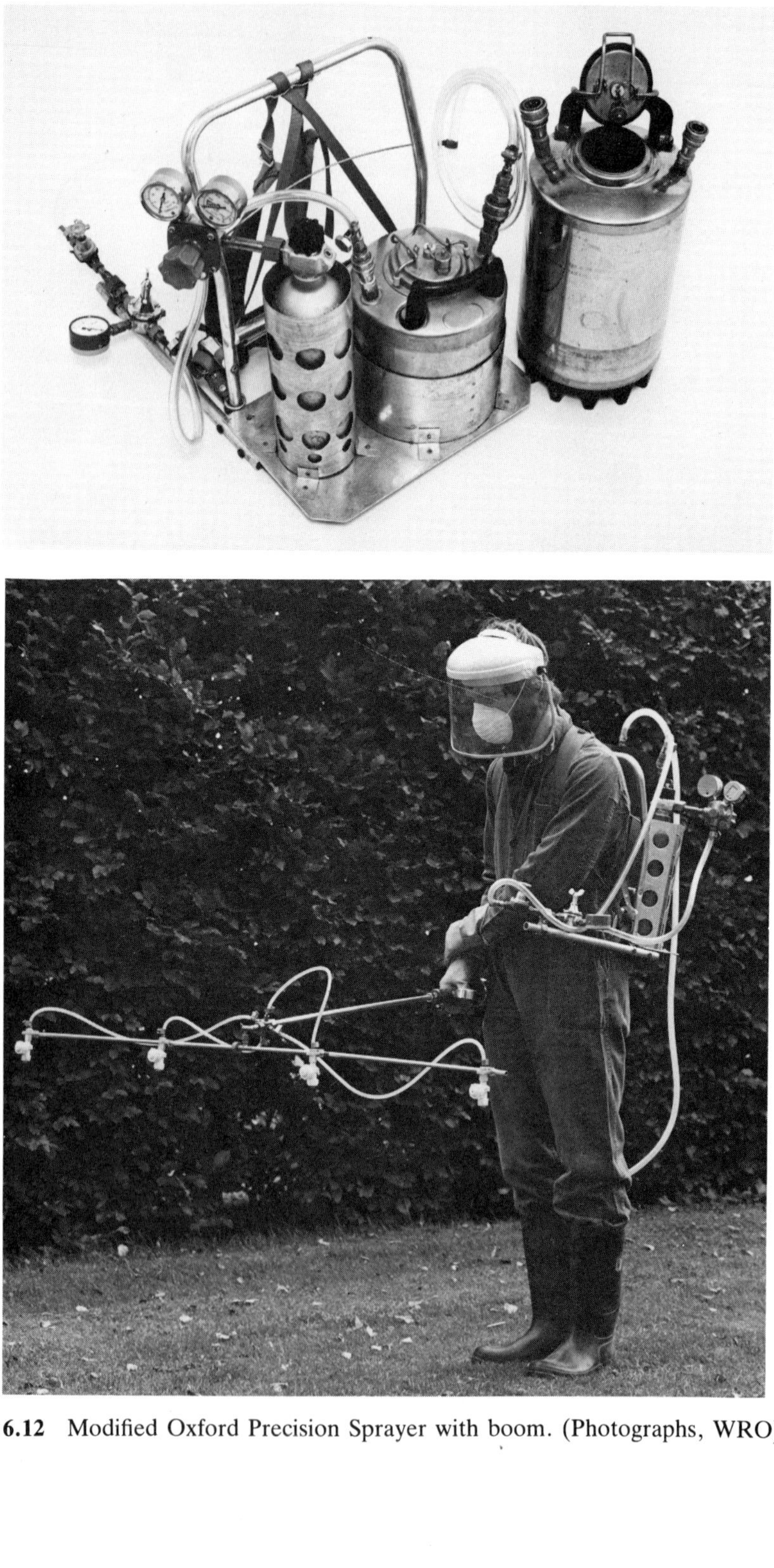

Fig. 6.12 Modified Oxford Precision Sprayer with boom. (Photographs, WRO)

Fig. 6.13 Shielded plot sprayer. (Photograph, WRO)

Anti-drip nozzles with variable spacing clamps are fixed to the lower section of the boom. At a volume rate of 250 litres/ha, 12 litres of spray will treat 480 m^2. One CO_2 cylinder can pressurize the 12-litre liquid container 17 times when used with the 4 m boom at an ambient temperature of 18 °C. Well over 100 plots can be treated in 1 day using a number of different pesticides at several dosages and sites. For transportation, the boom can be constructed in two halves which bolt together. Barrett (1974) modified the sprayer to apply herbicides to floating leaves of aquatic plants. An

Fig. 6.14 (a) Two-man portable boom
(b) Rear view of show back-pack on operator. (Photographs, WRO)

aluminium rig is supported by polystyrene floats and is towed by operators on each bank by means of guideropes. A third person on the rig controls the sprayer. Another portable boom has been described by Schopp and Landis (1961).

Recently a hand-carried boom was developed for electrodynamic spraying (Parham 1982) (Fig. 6.15).

Fig. 6.15 Two-man portable boom with electro-spray nozzles. (Photograph, ICI)

Wheeled plot-sprayers

Various designs of frame with one or more wheels have been designed to transport a sprayer with boom across plots (e.g. Rietz and Kohsiek 1982). Some have several spray tanks for rapid changing of the spray between plots. Many use sprayers similar to the OPS with propane, argon or CO_2 cylinders, but some have motorized pumps, or compressors. The height of the boom can be adjusted (Best and Wenhardt 1960).

The usefulness of small, wheeled, sprayers is somewhat limited, especially if wet soil or uneven ground makes it difficult to move at constant speeds. Small self-propelled sprayers often have insufficient power to overcome these difficulties.

(a)

(b)

(c)

(d)

Fig. 6.16 (a) Linear drive plot sprayer fitted to tractor
(b) Sprayer over crop while spraying
(c) Close-up of nozzle travelling over plot
(d) Controls of sprayer
(All photographs, ICI)

Mechanized small-plot sprayer

To achieve a more reproducible and even speed while spraying small plots, French (1980) has described tractor-mounted equipment to apply sprays within a 2 × 1 m shielded box constructed of light aluminium with a clear-domed PVC lid (Fig. 6.16). The box is straddled by three aluminium tubes, the 'rails' of the 'Mardrive linear transporter' which carries the spray nozzles over the crop. A compressor fitted to the tractor provides the compressed air to drive a 'mole' inside one of the tubes, and by magnetic coupling also the 'shuttle' to which the spray nozzles are fitted.

Compressed air is also fed to a stainless steel tank, designed to hold a 500 ml polythene bottle containing the pesticide or flushing fluid used to clean the spray lines. A gutter around the shield collects the flushing liquid and prevents any spray running down the shield. Pumps drain the waste liquid into a separate tank. The tractor allows rapid movement of the equipment between plots and provides the power to operate the sprayer; individual plots can be thus treated in less than 2 min. The speed of application across the pot and size of nozzles can be similar to field-sprayers.

Bradfield (1980) describes a self-propelled unit for the application of liquid and granular pesticides and fertilizers to turf areas. Baybee and Greulach (1980) also report a tricycle test plot sprayer transported to trial sites on a trailer. A vertical boom system for tree fruit and vines is described by Wallace (1979).

Apart from spraying pesticides, there is also interest in the application of insect eggs either to produce heavier infestations on experimental plots or to distribute predators. McWilliams (1979) describes apparatus to apply *Heliothis* spp. eggs suspended in water thickened with xanthan gum.

Mechanized sprayers for large plots

More rapid treatment of larger plots is possible with tractor-mounted sprayers adapted so as to allow rapid change of pesticides between plots. A compressed-air system is used on some sprayers to pressurize one or more liquid tanks via solenoid valve and regulator (Gebhardt 1972). The spray liquid is delivered to a manifold and thence to the individual nozzles. Time between plots is saved when several spray tanks are used to carry the different pesticides since this avoids having to clean the tanks during the trial. Nevertheless, the hosing has to be flushed between treatments. Rohde and Fay (1980) refer to a sprayer fitted with spray tank modules ranging from 2.4 to 100 litre capacity and Blanchard (1979) designed a high-clearance sprayer with six separate tanks. A standard commercial sprayer with a power take off driven pump is not ideal for small plots because of difficulties in cleaning the spray tank, pump and hoses. Some users have fitted an extra tank to carry water for flushing the system.

Electronic controls have been developed to monitor and control flow rate in relation to the forward speed to ensure the selected volume rate is applied, but further development of plot sprayers is needed. Comparatively little work has been done with equipment to apply ULV sprays on plots.

Taft *et al.* (1969) and Hare and Harrell (1978) experimented with ULV sprays using an auxiliary air supply to blow spray down on to the plants. Several sprayers have been designed to produce narrower droplet spectra than those obtained with hydraulic nozzles. Most of these have spinning discs. For experiments on herbicides a series of spinning discs was mounted on a vertical shaft and partially shrouded to provide a spray pattern equivalent to a fan nozzle. Five sets of nozzles were mounted so that spacing between them could be adjusted in relation to the swath achieved with different droplet sizes. Taylor, Merritt and Drinkwater (1976) describe a tractor-mounted version while Hind (1978) mounted similar units on a portable boom for small plots. Smith and Burt (1970) also shrouded spinning discs but added an airblower to assist removal of satellite droplets from within the shroud. Unshrouded rotary nozzles have been used by Burt, Smith and Lloyd (1966) to apply ULV sprays to cotton while Rutherford (1978) described a sprayer with three booms, one of which was fitted with spinning-disc nozzles to facilitate comparison with hydraulic nozzles on a field scale. Spinning cups have been mounted on a boom together with equipment to monitor both the flow rate and disc speeds during application, and air ducted over the boom from a fan can be used to create some turbulence to simulate aerial application (Fig. 6.17).

Garnett (1980) described a wheel-driven spinning disc which can be used for small-plot work, although droplet size is restricted to approximately

Fig. 6.17 Self-propelled plot sprayer with spinning disc nozzle and ducted airflow over nozzles (Photograph, Ciba Geigy)

250 μm. Small hand-carried battery-operated spinning-disc sprayers have been used in many herbicide and insecticide trials. Bals (1975a) described a herbicide sprayer to produce 250 μm droplets, while Matthews (1973) sprayed cotton with ULV insecticides using 50–100 μm droplets. On one sprayer, the number of batteries inserted in the handle can be adjusted so as to change the voltage and thus obtain different-sized droplets. The problems of plot size in relation to droplet distributions are discussed in the next chapter. Further studies are needed to assess the effectiveness of different droplet spectra, particularly smaller droplets, under field conditions.

Logarithmic sprayers

Most pesticides are applied at one dosage to a given plot. However, it is sometimes advantageous gradually to change the dosage rate across an area, thus allowing the minimum effective dose to be established (assuming that the target population is uniform). Similarly, dosage levels causing phytotoxicity can also be identified. Care must be taken to avoid drift, especially with small plots. The main disadvantage of logarithmic sprayers is that dosages are not randomized.

If a spray liquid is forced out of a container at a constant rate and replaced at the same rate with a diluent which is intimately mixed with it, the concentration of the sprayed liquid decreases exponentially. When the nozzle is moved at a constant speed, the dosage at any distance from the starting point can be calculated. The time taken for the dosage to reach half its initial value has to measured and related to the forward speed to determine the distance along the plot where the dosage has halved. Water is usually the diluent, though another pesticide spray liquid can be used when studying the effects of mixtures. Fryer (1956) first described a small-scale logarithmic sprayer which was then developed by Brunskill (1956, 1957) as a vehicle-mounted plot sprayer, but subsequently a portable shoulder-slung logarithmic sprayer was produced (Cialone and Bayer 1962; Clarke and Ross 1964). Other logarithmic sprayers have been reported by Day (1958), Leasure (1959), Yates and Ashton (1960) and Turner (1965). The calibration is normally checked colourmetrically by spraying a dye solution. Wright and Standifer (1969) also used a logarithmic sprayer attachment for a field sprayer.

Granule application

Wheatley and Niendorf (1969) describe a specially designed tractor-mounted multi-V-belt applicator designed to distribute small quantities of granular insecticides accurately along a required row length. The chemical or dose applied can be changed rapidly so that an experimental area can be treated plot-by-plot, rather than treatment-by-treatment. The equipment was originally designed to apply uniform doses along rows or broadcast the granules, but was adapted for logarithmically changing doses (Wheatley 1972). Subsequent work has involved the development of precision granule-metering belts to a precision seed-spacing drill, thus allowing accurate delivery and incorporation of the granules in the soil while metering seed (Thompson *et al*. 1981).

7

Is it effective?

Changes in agricultural practice are continually occurring as new varieties or economic pressures affect cropping programmes. These changes can affect pest populations. New management practices, therefore, need to be evaluated carefully under practical conditions before recommendations can be made. This involves field experimentation to determine the effectiveness of a particular treatment. Field experiments require considerable thought in the planning stage, so whenever possible the advice of a biometrician (statistician) should be sought early on to determine the most appropriate experimental design to obtain the maximum amount of the relevant information, particularly as resources may be very limited.

The biometrician will need to know the precise objectives of the proposed experiment, what treatments are planned and the percentage error in effects that can be tolerated. As much background information as possible will also be helpful in the discussions so that the biometrician is aware of similar work, what crop varieties, seed, sites and area of land are available to help decide on plot sizes, and sampling procedures to be used.

Field experiments may take months or even years to complete, especially with perennial crops, so supervision is very important throughout to ensure treatments have been correctly applied and the appropriate records taken. In any individual trial, the objective is to establish the significance of any differences observed between treatments. This means that sites have to be selected to provide as uniform an area as possible. As yields of adjacent similarly-treated plots can vary quite markedly in response to local differences in soil fertility, drainage, or some other factor (Fig. 7.1), all experimental comparisons must be replicated in space and preferably repeated in a number of years.

Sites should be sufficiently large to accommodate the trial without restrictions on plot size and number of replicates. A location representative of the

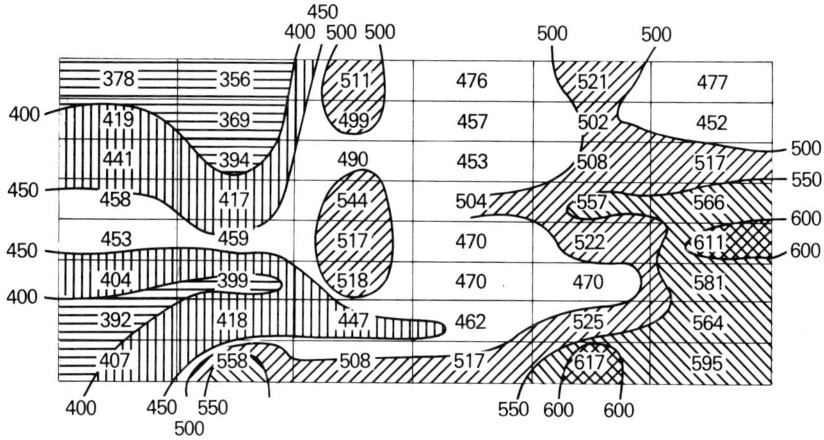

Fig. 7.1 Variation in yield of cabbages from a uniformity trial. (From Heath 1970)

surrounding area is needed, so that with adequate pest populations, significant effects due to treatments can be demonstrated. Easy access is important if frequent visits are made for recording pest populations, but adequate security is also needed to reduce the risk of theft of plot labels and crop. Where possible, a site should be chosen where there are good records of previous cropping practices.

Experimental design

Ideally, field trials should have the simplest possible design, especially when sampling of pest populations is needed. Some trials have such an excessively large number of treatments to include many different pesticidal products that few records are taken. These trials are often a waste of time and money, if any differences in yield cannot be explained adequately in relation to pest populations. The number of treatments in a field trial should be kept to a minimum by rigorous selection based on data from laboratory and small-scale trials as described in Chapter 5. If too many treatments are included, a larger trials area is needed and this increases the risk of greater variation due to differences between fields and other uncontrolled factors. This will reduce the accuracy of treatment comparisons.

Randomized block designs are often used in pesticide trials usually with from three to five replications depending on the number of treatments, the area available and difference to be detected (Table 7.1). Most trials have at least twenty-four plots so that there are not less than 10 deg. of freedom for estimating experimental error. Although laboratory data may indicate the dosage of a pesticide which should be examined in the field, persistence in relation to sunlight, rainfall or plant growth, for example, can influence

Table 7-1 Estimated percentage difference between two treatment areas that can be detected with 95 per cent confidence for varying number of replications and coefficients of variation (CV). (From Gomez 1972)

Replications	**Detectable difference (%)**		
	CV = 10%	**CV = 12%**	**CV = 14%**
2	22.6	27.1	31.7
3	17.2	20.6	24.0
4	14.5	17.4	20.3
5	12.9	15.4	18.0
6	11.6	14.0	16.3
7	10.7	12.9	15.0
8	10	12.0	14.0

its effectiveness, so field trials should examine more than one dosage level or different timings of application. Factorial experiments as used for different fertilizer levels are also appropriate for pesticide trials, for example, a factorial design was used when ULV treatments included different dosages, volume rates and swath widths in one trial (Matthews 1973a). An example of a factorial design (Fig. 7.2) is shown for a trial in which the effects of variety and fertilizer treatments on the control of pests are investigated. A complete factorial design should not be weakened by omission of any treatment-combinations unless they are clearly absured (e.g. if the concentration of chemical is too high). When cultural treatments are involved, a split-plot design may be more appropriate. Such trials may involve different row-spacings or cultivars in the split plots, the main plots receiving a spray treatment. Conversely, the main plots may receive different cultivation treatments and the split plots are designed to other variables like granule application. Choice of design will depend on where the emphasis is needed (i.e. whether on pest treatment or the cultural practices) and practical consideration such as drift of spray onto control areas.

A distinct advantage of randomized blocks of equal size is that experimental error is reduced while the flexibility and simplicity of a completely randomized design is retained. Each block can be orientated in a field in relation to any obvious fertilizer, shade, or soil moisture gradient or areas of weeds (Fig. 7.3), so in a field with a gradient along the length of the field, blocks should cross the field at 90 deg. to the gradient (Fig. 7.4). Blocks can be arranged to avoid obvious areas of poor fertility, drainage lines or other factors such as termite mounds which could affect the uniformity of the site. A map of each site showing permanent markers is essential for future reference so that as more information about a site is obtained, future experimental work can be planned more easily, and take account of any residual effects such as where certain herbicides or fertilizers were used. Wherever possible, experimental plots should be sited on fields which have had uniform crops for at least one season; using one year in a three-course rotation for trials is better than having trials with annual crops too frequently

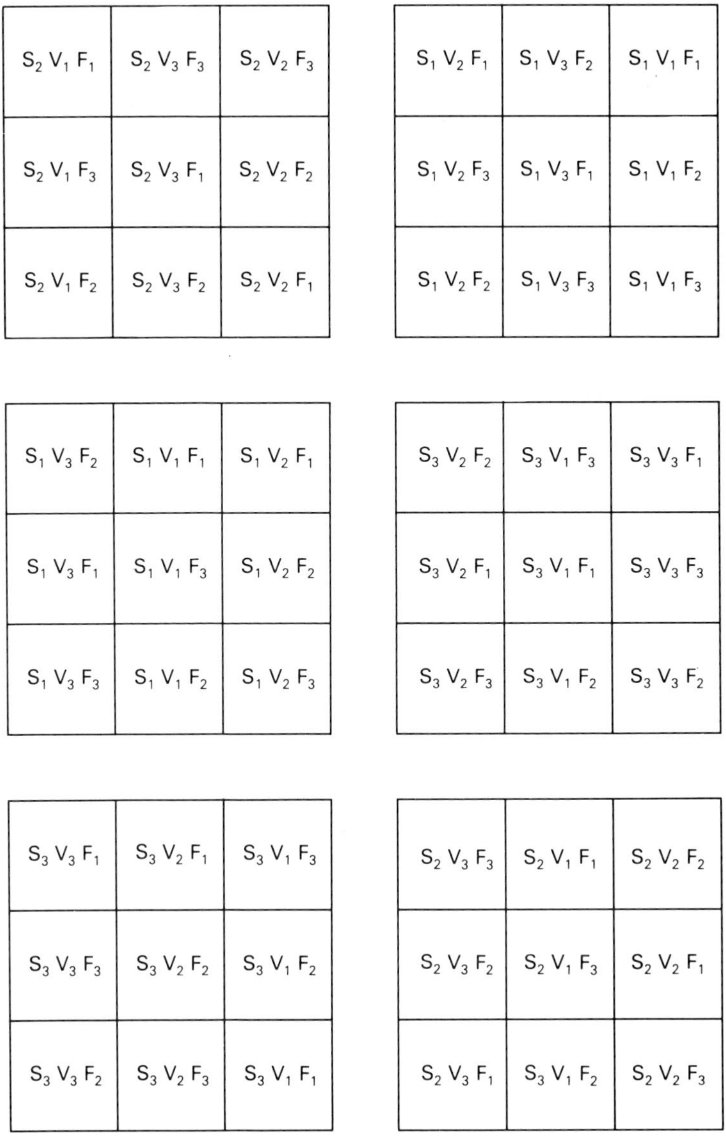

Fig. 7.2 A factorial experiment with pest control programme (S) on main plots. Different variety (V) shown as subplots and fertilizer treatment (F) as sub-subplot

on one site. When farm sites are involved, farmers who manage their field properly and are willing to co-operate need to be selected. This is particularly important with perennial crops.

When the fertility pattern is not known, blocks should be square (Fig. 7.5), or a latin square design used (Fig. 7.6). Latin squares are often better than randomized blocks when the number of treatments is small

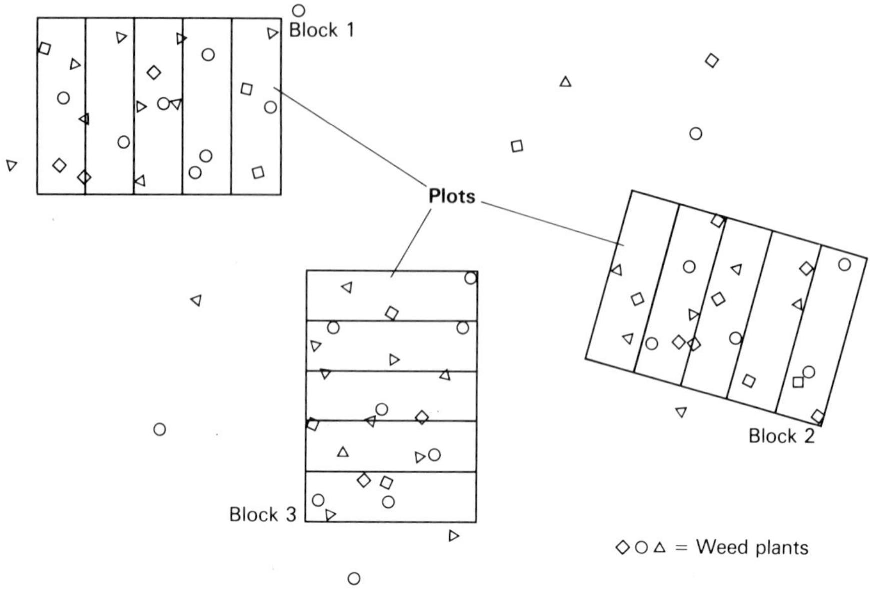

Fig. 7.3 Arrangement of blocks and plots on areas of field with known weed infestation

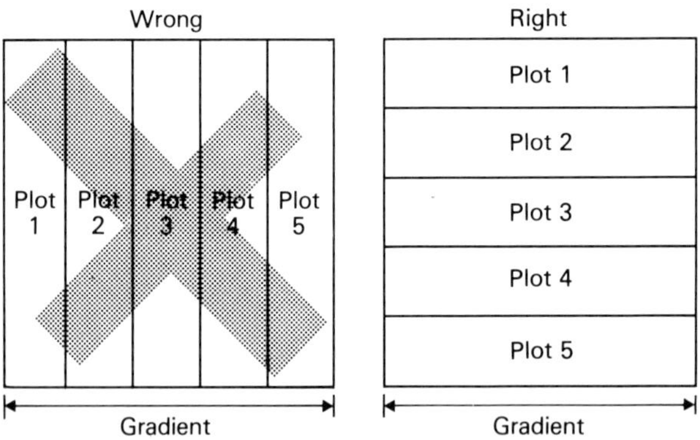

Fig. 7.4 Orientation of plots within a block to reduce the effect of a unidirectional fertility gradient

(4–7). Smaller squares do not give enough degrees of freedom unless the squares themselves are replicated. Larger squares are generally impractical as the number of replicates must equal the number of treatments. Two-directional blocking, so that each treatment occurs once in each row and once in each column, is useful as variability across a field in two directions can be factored-out in the analysis.

In plant-breeding trials where larger numbers of treatments are required, lattice designs with incomplete blocks are used. The number of plots in each

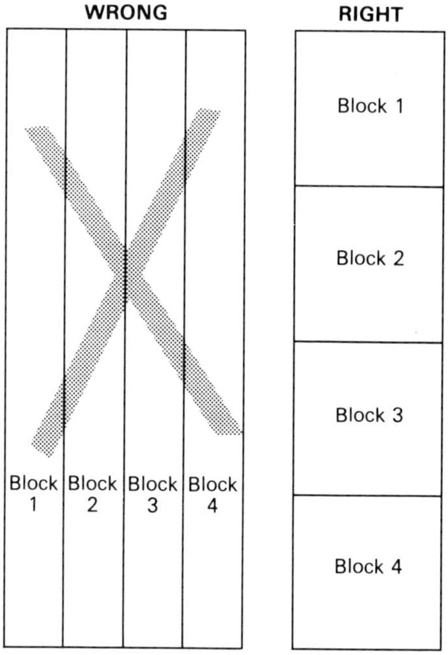

Fig. 7.5 Blocks arranged as square as possible when fertility pattern is not known

E	C	B	A	D
A	D	C	B	E
B	E	A	D	C
D	A	E	C	B
C	B	D	E	A

Fig. 7.6 Example of a latin-square layout with five treatments and five replicates

block is equal to the square root of the total number of treatments so the block size 8 would have 64 treatments in the trial (Gomez and Gomez 1976).

Trials on research stations may involve a large number of plots, but ultimately a treatment needs to be examined at many sites over a much wider area with plots located on farms. Such extension trials may have as few as two treatments at each site. Such trials are usually carried out to see whether a proposed treatment will be more economic for users in contrast to the existing recommendations. Tunstall and Matthews (1966) showed how an insecticide-spray programme could increase yields of seed cotton in Central Africa when compared with untreated crops. Subsequent trials did not include an unsprayed plot, but compared fertilized and unfertilized sprayed crops (Matthews 1972) and ULV vs. knapsack spraying (Matthews 1973b; Mowlam, Nyirenda and Tunstall 1975). There is considerable controversy regarding the use of untreated plots particularly in small-plot experiments. When an untreated plot is not present, there is no indication of how serious a pest, disease or weed was in the particular season in which the experiment was conducted. The other viewpoint is that untreated areas may provide a source of pests or disease inoculum which can spread over the whole trial area. Interference between plots in relation to airborne pathogens and pests is discussed by Jenkyn (1981) who prefers to keep total area of inoculum sources to a minimum. Reinfection of sprayed barley by mildew was decreased by widely separating the plots in a latin square, the intervening area being sown with a resistant variety (Bainbridge and Jenkyn 1976). In contrast, Lewis (1978) spread overwintered apple leaves from an abandoned orchard over experimental blocks to increase the infection by ascospores of apple scab.

This dilemma was referred to earlier in relation to assessing crop losses. Freeman (1967) discusses the problem and indicates some compromise solutions. Having an untreated plot in a separate area may not be very satisfactory due to variations in soil type and rainfall affecting plant growth and differences in pest incidence. Arranging untreated plots systematically throughout the trial so that each treated plot is directly compared with an adjacent plot is not satisfactory when spray treatments are involved, unless the plots are so small that they can be screened to reduce the risk of spray drift contaminating them. Semi-systematic designs with additional untreated plots have been used in conjunction with log-dose treatments applying granules for evaluating insecticides against some insect pests of vegetables (Thompson and Wheatley 1977) (Fig. 7.7). Assessment of disease level before the treatments allows use of covariance estimates of treatment effects, especially when unsprayed plots are not replicated throughout a trial site (Byrde 1981).

Untreated areas are often referred to as *control plots* although the pest is not being controlled on them; the term *check plot* is much better. In addition to untreated 'check' plots, some trials include the currently recommended treatment as a check, or plots sprayed with the diluent only to assess any possible effects due to the passage of a sprayer through the crop (e.g. soil compaction due to tractor) compared with completely untreated plots. Herbicide trials may include hand-weeded (or mechanically cultivated) and unweeded checks. An increase in the number of untreated plots to approximately the square root of the number of treatments in a trial improves estimates of the basic infestation and crop growth, but increases the labour involved (Thompson and Wheatley 1977). On some occasions, when the aim is to determine whether treatments are better than current practice, it has

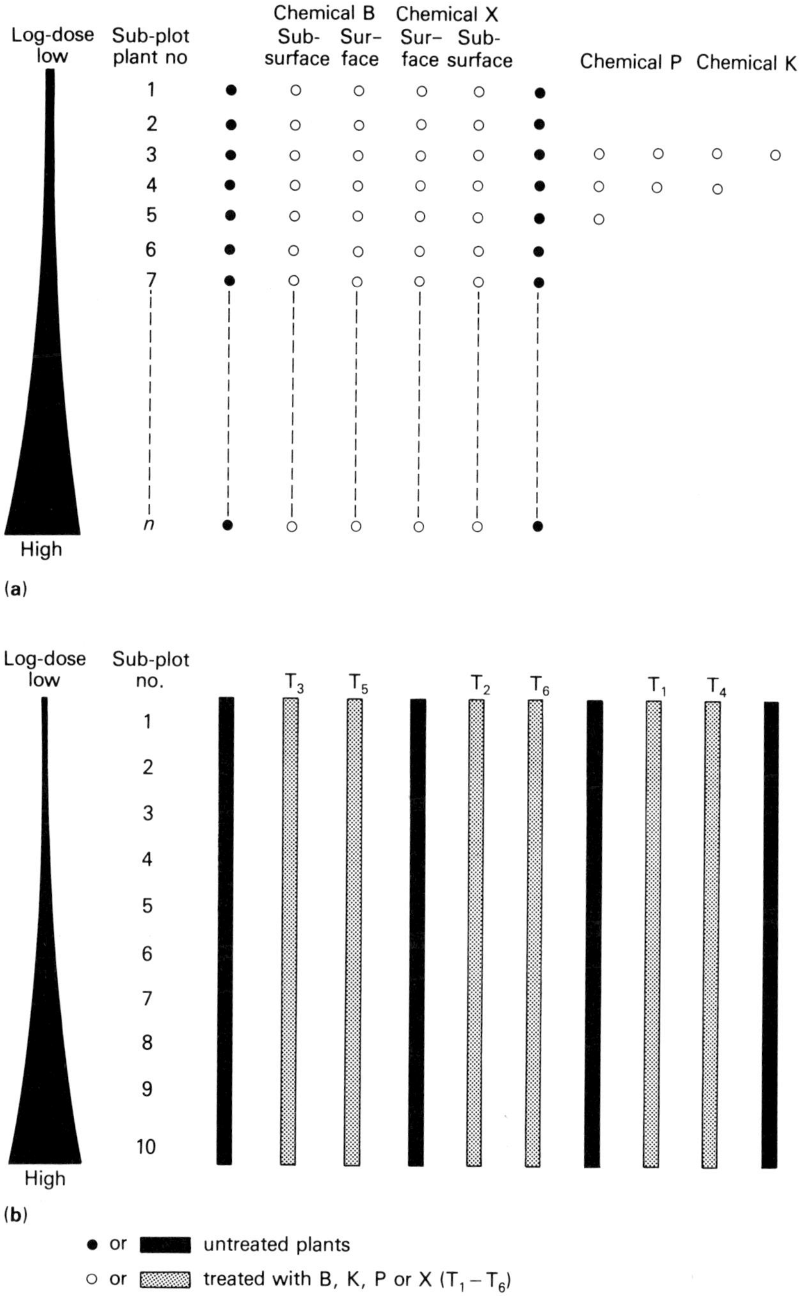

Fig. 7.7 Standardised semi-systematic design for log-dose insecticide trials for crop
(a) With plants spaced at low density in rows
(b) Sown continuously in rows at high plant densities

been suggested that there should be more replicates of the standard treatment (Pearce 1976).

Useful references include Heath (1970) who discusses the design of experiments, and Dyke (1974) provides a practical guide on how to do field experiments. Holly (1982) provides useful information on herbicide trials, while Ashby (1983) discusses the principles of insecticide and fungicide evaluation. Unterstenhofer (1963) refers to assessing the efficacy of insecticides. Statistical procedures are discussed by Gomez and Gomez (1976), Hammerton, Russell and Springer (1983), and by Pearce (1976) who refers to fruit trees and other perennial plants. Other useful references to experimental designs include standard statistical texts, such as Finney (1975), Bailey (1959) and Snedecor and Cochran (1967).

Plot size

Plot size can be selected to reduce the residual variance; thus Mercer and Hall (1911) concluded that little was to be gained by increasing the plot size of wheat beyond 80 m^2. Such a decision depends on results from trials with varying plot sizes, over several seasons on the same crop, pest and location. Too small a plot will increase the variance (Fig. 7.8), while larger plots use too much land. In practice, many other factors determine the plot size, and often the width of plots is dependent on the cutting width of a combine harvester. When tractor-mounted or aerial spray equipment is involved plots tend to be long and narrow, but this shape is not really appropriate in pest management studies. The increase in the edge affect is one defect, but where pesticide sprays are applied, the risk of contaminating adjacent plots is greater across narrow plots. Square plots are therefore preferred, especially as the wind direction is likely to vary during the treatments. Quite small plots can be used with fertilizer treatments or where granular pesticides or seed treatments are investigated, but plots treated with knapsack-sprayers generally need to be about 100 m^2 unless very small plots are surrounded by an effective shield during treatment. Plots measuring 30 × 30 m have been used with hand-carried spinning-disc sprayers, whereas a plot of at least 1 ha may be needed for larger equipment or aerial treatment. Allowance must be made for contamination of adjacent plots, so a small central sampling area is always needed. When tractor equipment has to pass through the experimental site several times, it is advisable to drive in the same direction along the same tracks each time to reduce mechanical damage to the crop, but the sampling area needs to be positioned to avoid effects due to the wheels. Plot size is also a problem when comparing different sowing or planting dates as it will be necessary to have access to individual plots for cultivation at different times. The movement of pests from early- to late-sown crops also has to be considered when interpreting the results. An experimental area may have to be surrounded by an non-experimental crop area to reduce edge effects or to maintain uniformity of cropping within a field for subsequent

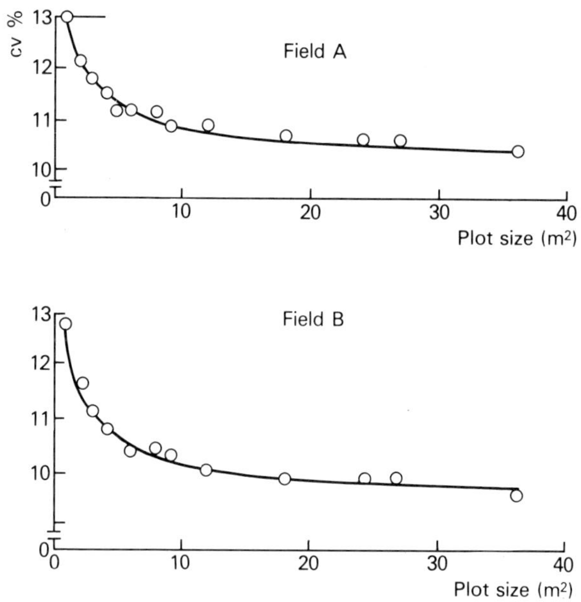

Fig. 7.8 Relation of coefficient of variation to plot size based on rice yield data from two uniformity trials

years. The area may be left untreated, but if it is an important source of re-infection or re-infestation of treated plots, then the non-experimental crop should receive the standard recommended treatment. Small plots may be needed when there is insufficient seed, materials, land or manpower, or there will be difficulties in disposing of harvested products contaminated with an experimental chemical, but results from such trials are often of very limited usefulness.

Extremely large plots are essential to assess the effects of pheromone traps and certain spray treatments, particularly those involving aerosols. Under these circumstances replication is virtually impossible, yet such treatments over an extensive community-wide area will have to be considered increasingly in a pest management programme.

Paths

Access to all the plots is needed for several reasons, so paths should be left between plots wherever possible. Inspection of plots and routine sampling of pests, diseases and weeds is much easier when the edges of plots are clearly demarcated without the need for strings. Paths are also essential when using knapsack-sprayers to allow turns at the end of the swath. Similarly, a path provides a stopping-place for a combine harvester between successive plots. Unless there are treatments at sowing or shortly afterwards, the crop can be sown over the whole field and the individual plots marked out later. The paths are then formed by cutting-out plants. A small motorized scythe has been used to create paths in cereals. Narrow 'tramlines' can be left unsown by blocking coulters on the seed drill. Plants alongside paths tend to get more nutrients and light, so they grow larger, but the effect is usually limited so sampling areas are not affected.

Sampling areas

The size of the sampling area relative to the whole plot will depend on a number of factors such as the type of crop or treatment (sprays, varieties, etc.), mobility of the pest and the records that need to be taken. When there is a highly mobile pest, the sampling area may have to be relatively small to minimize interplot effects (see p. 9) (Fig. 7.9). If frequent sampling is planned, then plots should not be too small, to reduce any effects caused by the movement of people through the area, and to allow for extra plants which have to be sampled. When harvesting is not mechanized, it is sometimes useful to remove the discarded plants alongside the sampling area so that supervision of labour is easier.

Guard rows or discard areas are also important near hedges or other barriers which can affect local air turbulence and influence the distribution of pests.

The yield area must be measured carefully to include half the space between rows on each side of the plot.

Randomization

Even on large-scale farm experiments which might involve only two plots, the treatments must be assigned at random. The flip of a coin can be used with only two treatments, but larger numbers can easily be drawn from tables of random numbers (see Appendix 2). When a randomized block design is used, if there are six treatments, you take six random numbers either vertically or horizontally across the table.

If the sequence is as follows:

1	298
2	671
3	442
4	038
5	630
6	096

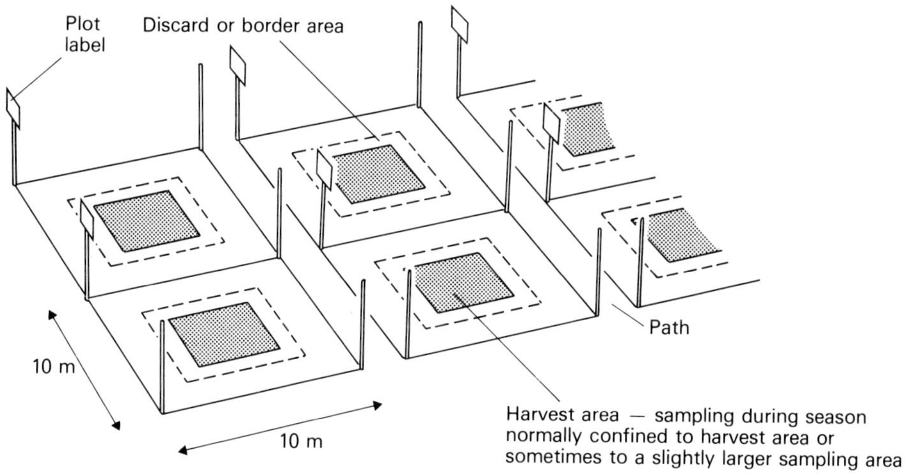

Fig. 7.9 Yield sampling area; discard areas and pathways in a field experiment

then the treatment numbers for the first replicate are decided by putting the random numbers in ranked order, i.e.
4, 6, 1, 3, 5, 2.
This is repeated for each block in the trial.

Random numbers can be obtained by picking numbers out of a hat, by using those reproduced in books of statistical tables or by computer programs to generate them. Sometimes when using random numbers, the plots of one treatment may be located adjacent to each other, e.g. if four blocks were arranged in a square. Similarly, one treatment may occur only at the outer corners. The experimenter should look closely at the proposed layout of the trial and repeat the randomization if particular sequences of treatments occur too often.

Marking out

When the design of the trial has been completed with the randomization of the plots, the trial has to be marked-out in the field. The site should have been examined in detail earlier, so that the experimenter will have a good idea of the position of the plots. If the crop has germinated before marking-out is necessary, patches of poor germination or other areas of poor growth can be avoided. If the crop is sown in rows, a base-line is established, preferably parallel to the rows, and along the longest line of the experimental layout in the field. The base-line is marked by two or more stakes or sighting poles. Ideally, the other three sides of the whole experimental area, or alternatively, of a single block, are marked-out. This will involve marking right angles. This can be done with a chain setting-out a triangle with sides in the ratio 3 : 4 : 5. This method requires at least three people to hold the chain and is very difficult to do without damaging seedlings, once the crop has germinated. The best method is to use an optical square, a small gadget with one or two pre-set prisms. The double optical square allows you to see to the left and to the right in the two prisms as well as directly forward through the intervening piece of glass. The optical square is held on the base-line by one person who directs a second person with a sighting pole until it is aligned with the poles seen in the prism. When the right angle has been fixed, it is best to put another line of stakes out and check the angle before measuring-out the individual plots.

A chain is better than a tape, which is difficult to keep straight if there is a strong side-wind. Wooden stakes or bamboo poles have been used as plot markers, but where termites occur, a metal marker is better. Plastic-coated curtain wire has been used, but when the crop grows tall, a more substantial metal rod is better. To guard against the possibility that plot markers are removed, the corners of the trial should be more permanently marked. A piece of metal with a wire can get entangled in a mower, so a single metal object may be buried so long as it can be relocated using a 'metal detector'.

Trials usually have square or rectangular plots, but this is not essential. Parallelogram-shaped plots have been used and when there are contour ridges, plots may consist of a number of rows each of a fixed length.

Plot labels may be marked in series 1, 2, 3 . . . but if sprays have to be applied to several treatments, it is better to have the treatment and replicate number clearly indicated. The two numbers can be differentiated by using roman numbers for the replicates, thus 2I, 2II 2III . . . (if a 3-digit number is used, the first digit refers to the block number and the remaining digits to plot numbers so 302 would be plot 2 in block 3). Different-coloured labels have also been used to assist spray operators treating the right plot, but actual treatments should not be indicated to avoid possible bias being introduced when the plots are evaluated.

In some trials, studying the effects of pesticides on natural enemies, plots were demarcated by a polythene barrier to reduce inter-plot movement of carabids (Edwards and Steed 1979; Shires 1980).

Treatment of plots

Equipment for the application of pesticides was considered in the previous chapter. The important point to remember here is that this equipment, as well as seed drills and fertilizer distributers used, must be carefully calibrated. The calibration should be made under 'field' conditions with the operators who will use the equipment in the trial. Several replicates of the calibration are needed to check its accuracy. If different types of granules are applied, a calibration with each formulation is needed. When a number of different treatments are applied with the same equipment, great care must be taken to clean it properly between treatments. Washing-out a sprayer should include pumping clean water through the nozzles, and not just rinsing-out the spray tank. When applying a herbicide at several dilutions, it is best to begin with the lowest dilution. If different formulations are in a trial, solutions should be applied first, then wettable powders and, finally, emulsions. Normally wettable powders and emulsions are not mixed, but if a small quantity of an emulsifiable concentrate has to be added, the wettable powder should be mixed and diluted first before the emulsifiable concentrate is added. Great care should be taken to avoid mixing too much chemical. Very small amounts of excess diluted spray and washings from the sprayer can be poured into a small hole dug near the trial. Formulations which do not mix with water will have to be returned to the chemical store and disposed of later in accordance to local regulations.

All the treatments in each block should be completed before moving on to the other blocks. This is very difficult when pesticides are applied unless there is sufficient equipment or separate spray tanks for each treatment to minimize the time needed for cleaning. The total time required for treating a trial should be as short as possible to avoid treatments being applied under different weather conditions. In the tropics the temperature can change considerably between 06 00 and 10 00 hrs. Similarly, wind strength or direction and RH can vary, so a simple anemometer and whirling hygrometer should be used to take records over the period of treatment. The period of time before rain following an application should also be noted as rainfall may reduce the effect of some treatments.

Manual weeding of plots should be performed carefully to avoid damage to the crop. Several weedings may be necessary on a check plot, but excessive cultivation could have a detrimental effect, e.g. by soil erosion. Some crops, such as groundnuts, are particularly difficult to weed by hand, because pulling out perennial grasses disturbs the crop.

Recording data

When planning a trial, the experimenter usually decides which records should be taken and at what frequency, so that any assistants required can be trained and their duties allocated. Nevertheless, regular visits are essential to make general observations and 'assess' progress of the trial. Important information may be obtained that would otherwise have been missed by the routine records.

Methods of recording experimental data in the laboratory and field are rapidly changing and use of standard 80-column punch cards, will soon be out of date (Berry *et al.* 1966). Rowell (1971) used similar cards marked with a pencil and read by an optical mark reader to transfer the data directly to a computer. Accuracy was verified by checking a printed list from the computer against written figures at the top of the marked cards. Recent developments with portable data terminals allow data to be keyed directly into a memory store or magnetic-tape store. Alternatively, data can be transmitted via a normal telephone line to a computer. The terminal can also be connected to a suitable printer to allow the field assistant to check what data he has collected. Less expensive computer systems will undoubtedly lead to greater use of more rapid methods of transferring data, but in the majority of experiments, data will continue to be recorded initially in a notebook or on to specially designed forms. Copying data must, however, be kept to a minimum. Errors often occur and all the data must be checked before analysis.

Subjective assessments are often necessary to examine a large number of plots quickly. Counting mites is difficult (see p. 26) when populations are large, so scoring is used. Similarly, scoring systems are frequently used to assess weed populations, and crop injury levels. When subjective counts are used, two people ideally evaluate a trial independently and their results are pooled. Where possible it is much better to have quantitative measurements. Some of the techniques described in Chapter 2 can be used in experimental plots. Collecting quantitative data takes time and can be expensive, so sampling should be planned to provide as much information as possible in relation to the available resources. When assessing weed control, the actual density of weeds measured using quadrats is very useful. In some experiments, the weeds can be harvested to determine fresh and dry weights. When removal of plants is involved, sampling is best restricted to an area surrounding the harvest area (Fig. 7.9, p. 157) to avoid affecting yield data. In some cases plants can be removed to measure growth without appreciable effect on yield. Uprooting of cotton plants enabled neighbouring plants to grow larger and compensate, but removal of the same number from each plot should have a proportional effect even if no compensation occurs. Measurement of plant growth in relation to pest infestation is very important as economic thresholds are likely to vary according to the stage of plant development.

Apart from measuring effects on pests, an assessment of beneficial insects is also needed. One method of evaluating the effect of foliar sprays on egg parasites is to grow susceptible crop varieties in pots, cage the pest over each pot for 24 hr and then put the plants with eggs of the pest in the experimental plots. After exposure for two or three days, the pots are taken back to the glasshouse and samples are transferred to glass tubes plugged with cotton to assess egg and parasite hatch. This technique has been used to assess parasitism of brown planthopper on rice.

Whatever samples are taken from individual plots, the position of the samples within the plots must be at random. Random numbers can be given to tell the recorder which row, and how far along the row he should walk, before taking records.

Farmers are concerned with the economic return from crops, so most field trials are carried through to harvest. The yield area should be marked-out with stakes and surrounded by a string, particularly if plots are harvested by hand. On cereals small combine harvesters are available, but it is very difficult to empty the machine completely at the end of each plot, in which case an unknown quantity of grain is carried over from each plot to the next. Timing of harvest is often important, as any delay may cause an increase in losses, e.g. by birds or plants shedding grain. Harvested produce is usually taken in some type of container to a central point for weighing and grading. As labels fixed to the outside of a container are liable to get torn off, a separate label should be inside it as well. It may not be possible to weigh the produce from all the plots on the same day as harvesting; therefore, it is important to double-check that all the containers have been brought-in from the field and safely stored. If a quick check reveals missing containers, they may be found if a search is made immediately to ascertain whether they had been accidentally left in the field. The yield from all the plots should be in the same size container to allow the balance to be tared. Spring balances are often used as they can be hung from a beam or tripod, but their accuracy needs to be checked regularly. When weighing sacks of produce from large plots, a butcher's scale-yard is robust and easy to use. Direct recording balances are now available and can produce a record of weights suitable for transfer to a computer. When these balances are used, an immediate print-out is needed to check for obvious errors or missing data while the produce of the individual plots is available for reweighing. When weighing yield data, two people should be present, with one to record and check the other person's weighing. When the trial has been harvested, it is not possible to return and take another sample if one is lost, so great care is needed to ensure accurate data. Records of weighings should be duplicated using a carbon paper or carbon-backed paper, with one copy filed for safety in case the top copy is lost. If a weight is recorded initially incorrect, then alterations should be seen clearly by crossing-out the wrong figure and writing the correct figure above it. As with all records, copying should be avoided, but when it is necessary, always get someone to verify that the figures have been copied correctly.

When weighing the produce has been completed, it is sometimes necessary to take a sample for further investigation. Samples of grain may be needed to determine the moisture content, seed cotton for ginning percentage or lint quality; another example is the analysis of sugar content in beet or cane.

Analysis of results

There is now a wide range of computer programs for different methods of statistical analysis of data. The danger is that numbers will be analysed without a 'real feeling' for the data. The experimenter should examine his data very carefully before they are analysed and 'sniff out' any unusual results. A biometrician on one occasion noticed a very unusual figure in a set of yields. On investigation he found that the plot number had been added to the yield!

When an analysis of variance has been completed and an error mean square determined, the experimenter usually makes a number of comparisons to assess the significance of differences between treatment means. An observed difference is significant at the 5 per cent level when such a difference, or larger, would arise by chance only on 5 per cent of occasions if there was no real difference. Many authors refer to the least significant difference (LSD) (Snedecor and Cochran 1967), but this test is only correct if it is applied to a randomly selected pair of means. In practice, the experimenter wishes to compare pairs which are not randomly selected. The alternative is to use a multiple-range test, based on the number of means which are compared. This is considered less likely to show differences that do not really exist. Duncan's multiple-range test is the most commonly adopted (Duncan 1955; Parker 1973). Nevertheless, there may be very few situations which require such tests to separate means into significantly different groups, as all the comparisons may not have equal importance. Some comparisons might be pointless where the differences obviously exist (e.g. between untreated and a number of higher-yielding treated plots). As no definite rules can be given about what comparisons are valid, judgement is needed in relation to the original objectives of the experiment. This emphasizes the point made earlier, that the experimenter should consult a biometrician at the planning stage and together decide what comparisons should be made.

Regression and correlation analysis can be very useful, e.g. when establishing if there is a relationship between the level of pest incidence and yield, or whether there is any crop compensation. Yield will undoubtedly be influenced by several factors, but it is easier to understand if only one independent variable is studied at a time. A simple regression analysis can be used when data is obtained from a controlled experiment in which as many factors are kept constant while only one or two factors under investigation are varied. The relationship is expressed generally by the equation $y = a + bx$ where y is the dependent variable, such as level of pest incidence, and x is the independent variable such as yield; a and b are constants. Chiang (1973), for example, showed a simple linear relationship between per cent yield reduction and the number of northern corn rootworm (*Diabrotica longicornia*) larvae for the maize plant. Teng (1981) refers to the use of regression analysis for developing crop-loss models. When losses commonly expressed as a percentage are estimated from a single assessment of pest or disease at one growth stage, he refers to critical point models. When assessments are made sequentially at progressively more advanced growth stages a multiple-point model is used. When more than one variable is involved, as in factorial experiments, multiple linear regression techniques are required. However, most relationships in pest management are not linear and an appropriate non-linear model has to be determined, e.g. polynomial, sigmoid, logarithmic or exponential functions.

Presentation of results

The results will have to be presented in different formats bearing in mind the requirements of the readers. Farmers will be more interested in the financial implications of the results, so cash returns will be appreciated more than the detailed statistics quoted in scientific papers (Clarke 1965). Freeman (1981) also advocates avoiding too much numerical information, pointing out that statistics are like scaffolding: 'When a building is under construction, it needs support, but when it is finished, it should stand up by itself.' Reports of experiments should include information on its location and previous history, together with overall treatments such as variety, seed rate, fertilizer and cultivations. Dimensions of the plots and methods of sampling should be given with technical details of the individual treatments. Tabular data should be clear without reference to the text. Tables which are too large will fail to provide a readily understood summary of the data. Usually a graphical presentation is easier to understand. Ehrenberg (1978) discusses the advantage of graphs to communicate the qualitative aspect of data.

In the tables, figures should not present more digits than are strictly necessary; thus data should be rounded-off in relation to their accuracy. It has been suggested that to avoid bias when rounding-off 5s, the nearest even number is used whether you round up or down. Yields are often converted to kg/ha, but when very small plots are involved it is much better to use g/m^2, or g/plant. A concise summary of the main results is perhaps the most important part of a report, so that the relevant information can be noted by those too busy to read all the details in a report.

Data on individual replicates is not usually required unless treatment × replicate interactions are significant. When several sites are involved; an overall analysis should be carried out to study treatment × site interaction. In this case, treatment means are normally reported for each site.

Presentation of results obtained from factorial experiments will depend on the interactions revealed (Ridgman 1969). If there are significant interactions, means for the main effects will not be of much value. If data have been transformed prior to analysis, the means of the original data should be presented alongside the transformed means.

A report should include standard errors with the means to allow readers to use their judgement in assessing the value of the results in the light of their own experience.

Operational problems

The success of field experiments depends initially on very careful planning. Many people will inevitably be involved and transport will have to be arranged. On research stations, the experiments of individual scientists have to be fitted into a limited area and financial constraints will also influence their final layout. Early planning is essential as equipment or materials may have to be specially ordered and if this necessitates transport from another country, delays in shipment, customs clearance and other formalities must be considered. After the initial planning and requirements have been ordered, staff may have to be trained, and in the case of farm-scale trials, the proposed sites have to be visited and the proposed trial discussed with the co-operating farmers. When the site has been chosen it should be mapped. Regular liaison visits are needed throughout the trial to ensure the smooth running of the trial. In large-scale trials, good communication between all those participating is often of crucial importance.

This chapter has been primarily concerned with field experimentation needed to sort out whether a particular treatment has an effect on a pest population. Increasingly there will be a need to examine various systems of managing pest populations over a large area. This will require a different approach to the formalized field trials. The aim will be to determine whether particular combinations of tactics fit together to provide a coherent strategy which is cost-effective. This will necessitate an evaluation of existing management practices and determination of the resources available to understand the constraints and managerial limitations which have to be overcome before improved technology can be adopted. Close liaison with extension services will be essential so that economic improvements can be introduced and assessed on a large scale before being formally recommended.

8

Pest management

Considerable attention has been given in recent years to 'pest management' or 'integrated pest management' (IPM) following increasing recognition that sole reliance on one method of control, particularly involving the use of pesticides, is counterproductive. Following the introduction of DDT, some entomologists such as Ripper (1944) pointed out that insecticides would also kill natural enemies and result in resurgences of pest populations. Later Carson (1962) focused attention on the broader implications of unrestrained use of pesticides. But probably the most significant factor encouraging a return to more ecologically orientated methods of controlling pest populations has been the development of pesticide resistance in pests. Selection of genetically resistant populations can occur quite rapidly under certain circumstances and is already a major problem limiting the use of certain insecticides and fungicides. There is also now resistance in some weeds to herbicides which previously killed them.

Before the widespread use of organic pesticides began in the 1940s, man did integrate various control strategies. This occurred sometimes on a national scale, e.g. the use of varieties resistant to jassids and closed seasons enabled cotton to be established in southern Africa. The return to integrated pest management now is strengthened by the availability of new 'tools'. There is little doubt that pesticides will continue to provide a valuable weapon in pest management programmes if used wisely and more precisely (Southwood 1979).

Also, the microprocessor will be increasingly important in pest management programmes, and will enable more rapid analysis of data, which will facilitate decision making.

Today, pest management is much more than simply integrating the use of two or more control tactics. It is concerned with the combination of appropriate strategies and tactics to optimize net benefits of pest control throughout a given area or ecosystem. The aim should be to provide a system which is sufficiently robust to maintain control over a prolonged period.

In the agricultural sector, pest management programmes embrace the whole area of crop production, including marketing, and require careful analysis of the complex biological and economic interrelationships. To gain an understanding of the complexity of the problem, systems analysis is a valuable research tool providing a means of describing the biological, technical and managerial factors involved. This analysis is mainly concerned with identification of key questions to assist with making both research and management decisions, but it can be extended with existing field data to construct computer-simulated models of pest populations and crop development. These can be used to assess various pest management strategies under a variety of conditions. The value of such a model should not depend on whether it gives an exact answer, but whether the gain in understanding, predictive ability and management efficiency exceeds that which would have been obtained without using a model (Shoemaker 1980). Apart from the improvement in understanding the biological/ecological situation, systems analysis also needs to be judged by its impact on improving pest control. The systems approach has been widely adopted in North America to examine important crops, such as cotton (Gutierrez *et al.* 1980), soyabean (Rudd *et al.* 1980), alfalfa (Ruesink *et al.* 1980), deciduous fruit (Asquith *et al.* 1980) and citrus (Luck, Allen and Baasch 1980). So far, much of the effort has been to develop simulation models which interlink the growth of the crop and the effects of pests, but attempts are being made to translate research results into computerized systems for use by extension staff. In Texas, the pilot system known as BUGNET is intended to provide farmers with the latest information affecting management decisions (Hartstack 1982). Similarly, a computerized decision-making program SIRATAC is used in Australia (Wilson 1982). Elsewhere, Conway *et al.* (1975) examined the control of sugar cane froghopper to optimize the spraying strategy. In Japan, Kiritani (1977) modelling the green rice leafhopper, showed the importance of natural enemies, the possible integrated use of spiders with a moderately resistant variety of rice, and the value of cultural controls, especially winter ploughing of fallow paddy fields, to suppress the incidence of rice dwarf virus transmitted by the leafhopper. Prediction of weed infestations and the cost of differing control strategies has also been examined by computer modelling (Cussans and Moss 1982), and Curtis, Cook and Wood (1978) used a model to examine suppression of resistance in anopheline mosquitoes. Ruesiek (1975) gives a general account of the analysis and modelling in pest management.

Management decisions

Pest management decisions have to be taken by those directly concerned with controlling pests, e.g. farmers, those allocating resources for research and advisory work, those providing advice and the government itself in relation to regulatory matters involving environmental protection.

Decision-making by the farmer

The way in which a farmer decides which strategy or tactic to adopt to control a pest is influenced by three factors (Norton 1977): 1, his perception of pest attack and damage; 2, the range of action available to him; 3, his objectives, including such criteria as profitability, reliability, convenience and safety.

There have been few studies of the farmer's perception of pest attack and damage. In Eastern England, Mumford (1977) found that most sugar-beet growers were able to estimate average losses reasonably well, when these were compared with field trial data, but often overestimated the worst losses.

The range of methods available for a particular farmer will depend on constraints such as availability or sufficient finance for appropriate machinery, chemicals or labour. Lack of information and training may deter acceptance of a new technique. In some areas inadequate distribution may prevent a farmer purchasing the equipment or chemicals he needs. Government regulations can also affect the choice of control strategy.

The farmer is therefore continually faced with making decisions. The options available to a farmer can be represented by a decision tree (Fig. 8.1). As the crop matures, the range of options is reduced. His decisions will be determined to a large extent by the relative cost and performance of each option, but many farmers are 'risk averse', i.e. they may well adopt a less profitable option when it is less risky (Norton 1977). Potato-growers, for example, are more likely to spray a fungicide to control blight, rather than risk the slightest chance of making a loss. Unfortunately, the farmer may have only a vague idea of the probability of the result of each strategy available to him, so his decision is also influenced by his own or his neighbour's experience gained, very often, by trial and error.

In certain situations, pest management decisions can be improved if the farmer has access to more information. One important part of his information is the level of pest infestation and what damage it will cause at various stages of crop development. Crop monitoring is one method the farmer can use himself, if the extension services can provide training in methodology and interpretation of the results. In Zimbabwe, scouting of cotton crops is supported by a levy for the training programme (Burgess 1983). In some cotton-growing areas in the USA, crop-monitoring data from individual farms within a community can be processed by computer to allow a rapid feedback to advise whether or not control action is needed (Phillips, Gutierrez and Adkisson 1980; Rijsdijk and Zadoks, 1979; Zadoks 1981; Rijsdijk 1982). Meteorological data is also used particularly in disease-forecasting (Thompson 1982) Each farmer can then base his decisions on a much larger pool of data, and relate his own local situation to the general trend

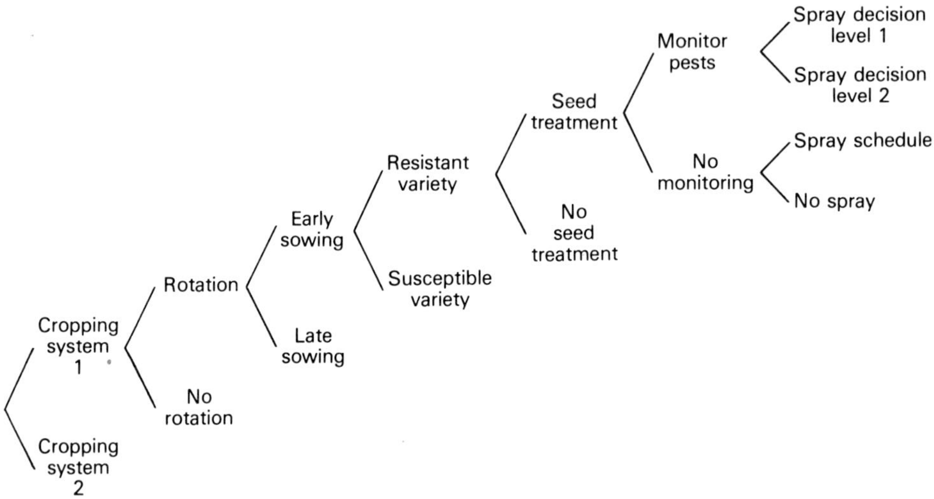

of pest infestation. In the UK, information on aphid infestations has been posted to sugar beet growers. Local radio networks can distribute such information in a similar manner to routine weather forecasts, but it is possible that there will be a trend towards using teletext systems linked to a computer, so that farmers can interact with a computer database on which the cropping history and pest infestation records for individual fields can be stored. While it might be argued that a farmer can keep his records in a notebook, progress in microprocessor design suggests that if farmers were able to store much of the data related to their own farms in a computer which is interlinked with a larger regional database, decision-making could be improved. More general information, including commodity prices, recommendations for different pest situations and the overall pest situation, would be available on a regional basis. In the UK, some information is available through CEEFAX and PRESTEL the latter operated via a telephone link, and a pilot scheme to store the records from individual fields on a central computer has started. When farmers have confidence in data provided by such information retrieval systems, particularly in relation to forecasts of crop loss and recommendations, there should be less need for routine use of

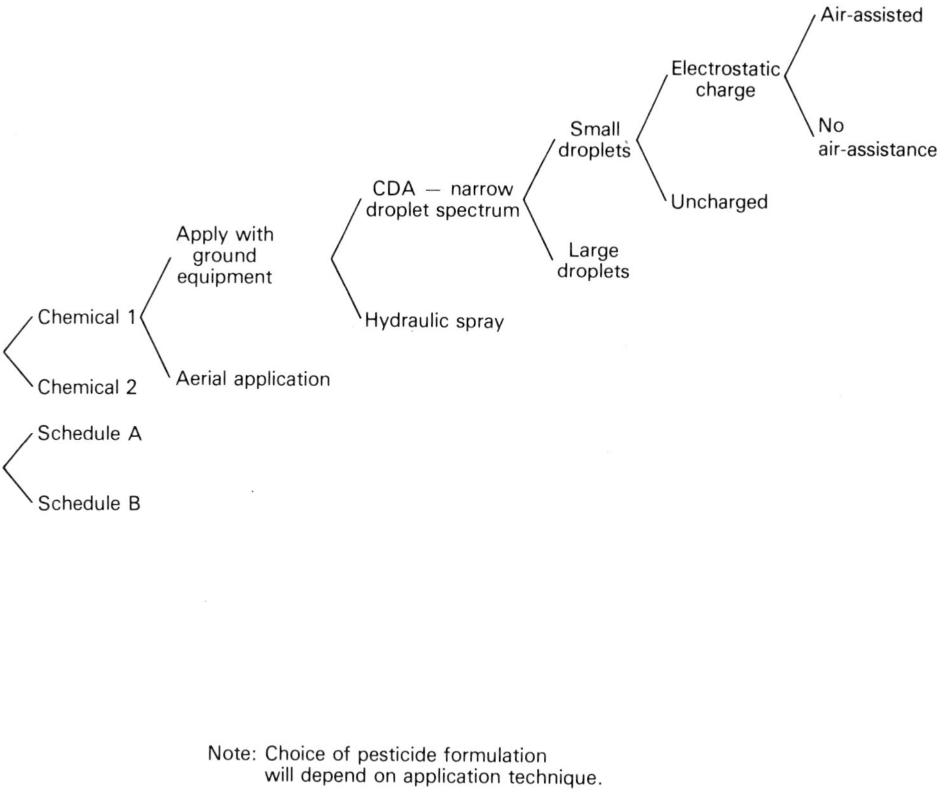

Fig. 8.1 A decision tree for integrated control. (Adapted from Norton 1977)

pesticides and greater reliance on treatment only when necessary according to economic thresholds. The farmer will require more training to understand pest management decisions, or he will rely more on specialist consultants with experience to help interpret data and help make such decisions.

Consultants already provide farmers with useful advice, but primarily of an agronomic nature as they are concerned with effective use of fertilizers and pesticides. Their role is likely to increase and the IPM specialist will have a much broader outlook to appreciate the implications of different control tactics and thus will have had more biological and ecological training.

Whether the farmer seeks advice from a consultant or not, those who already rely on pesticides may have to modify their attitudes to decision-making. In the USA, reluctance to change to an IPM programme has been due to growers not being accustomed to: 1, use of economic thresholds; 2, regular monitoring instead of random or irregular checks; 3, making decisions in a dynamic context, i.e. considering crop stage, natural enemies, weather, etc. instead of a fixed routine programme which is easier to operate; 4, using controls other than pesticides; 5, the adoption of a community-wide programme (Flint and van de Bosch 1981).

Table 8.1 Average cost ($) of insecticides per hectare 1970–71. (Based on Hall, Norgard and True 1975)

	Users of consultants	Non-users of consultants
	$	$
Cotton	4.94	11.97
Citrus	20.53	42.35

However, fixed programmes are getting more expensive relative to crop prices, so growers who have used IPM consultants in California have been able to reduce their cost of insect control on their cotton and citrus crops even when consultancy fees are included (Hall, Norgard and True 1975). (Table 8.1).

Practical pest management systems have been developed step-by-step by modifications to existing pest control programmes as experimental data becomes available.

More information is clearly needed to understand the complexity of crop–pest interactions, but as Way (1973) pointed out, we should not exclude simple answers. Careful observation is of critical importance. The planting of blackberries to provide an overwintering host of a crucially important parasite of the grape leafhopper followed the observation that the leafhopper did little damage in places where blackberries grew wild. Similarly, integrated control of phytophagous mites in apple orchards was possible following the observation that the mite predator *Typhlodromus occidentalis* was resistant to organophosphate acaricides in North-West USA (Wildbolz and Meier 1973). Thus resistant natural enemies can easily survive the careful use of chemical control.

Appropriate research is clearly needed to provide background information and improve IPM programmes. The systems analysis approach referred to earlier is often very relevant to indicate which important gaps in our knowledge should receive priority for funding research. A description of the problem is required initially. Norton (1979) illustrates one part of this using the situation found in an apple orchard. To indicate the stage of each pest in relation to plant development a time profile is used (Table 8.2). Too often one pest is considered in isolation, control of which may allow scope for another to take its place, but this time-profile shows the diversity of pests on a single crop, their development over time and their location in the agro-ecosystem.

Damage caused to the components of the tree by these pests is illustrated in a damage matrix (Table 8.3). Certain treatments have a detrimental effect on the beneficial predators (Table 8.4). As fungicide applications are essential for mildew control, choice of fungicide to reduce interference in the interaction between mites and their predators is very important, to avoid having to rely more heavily on acaricides to control mites. Thus, binapacryl supresses spider mites and also nearly eliminates the predatory mite

Typhlodromus pyri, whereas bupirimate or triadimefon are non-toxic to mites. Cranham and Solomon (1981) list pesticides which are harmless or have a low toxicity to *T. pyri* (Table 8.5). As indicated earlier (Ch. 4) assessments of the impact of any pesticide, including fungicides and herbicides, on beneficial or non-target organisms is clearly important in improving advice to farmers.

Apart from choice of pesticide, the farmer needs better information on when to apply it. The empirical approach is needed, e.g. trials on cotton investigated the effects of sprays at different thresholds based on egg counts (Matthews and Tunstall 1968). Conway *et al.* (1975) used a model to examine the effect of different spray strategies on a single froghopper brood. The most profitable strategy for a range of brood sizes is shown by the arc subtended by the lines in Fig. 8.2. The situation is more complex when the different broods during the season are considered together with changes in froghopper populations in relation to their density. The systems analysis approach enables a more detailed examination of the possible consequences of different control strategies, but the result invariably depends on a number of assumptions. Like all models, their acceptability will depend on how well the data relates to the real world. The main result of some modelling will be to define and select the treatments which should be examined in the field, but suitable field trials may not always be possible due to constraints of time, finance or trained personnel. The use of a model can be in these circumstances the only way of making a recommendation for farmers. Models of the cattle tick *Boophilus microplus* in Australia have been used to search systematically for an alternative robust control strategy to avoid reliance on chemical control. In this case, Norton, Sutherst and Maywald (1982) examine the management options open to cattlemen (Fig. 8.3). Improving the herd's resistance to the ticks by breeding and by culling susceptible animals is regarded as essential, but this has to be integrated with control of movement of cattle between paddocks and an improved dipping strategy such as dipping in late autumn and early winter when low temperatures prevent the ticks which survive the dipping from producing viable progeny.

Removal of susceptible individuals from a population by pesticide application will inevitably result in the selection of a resistant population. How quickly a population will become resistant and over what area, depends on the selection pressure and genetic background of a given pest population. An area of multidisciplinary research which is of paramount importance is concerned with developing ways of countering the problem of resistance to pesticides (Sawicki 1981).

The effectiveness of existing pesticides needs to be conserved by developing control strategies which prevent or delay the onset of resistance in pest populations (Sawicki 1981). Computer studies by Comins (1977a, b, c) and others have indicated some of the options available to delay resistance (Denholm 1981). These options include improved timing of pesticide applications to keep the number of applications to a minimum and thus reduce selection pressure for resistance. An increase in pest populations is usually

Table 8.2 Time profile for components of the orchard pest system. (From Norton 1979.) Stages of blossom development are bud-burst (BB), green cluster (GC), pink bud (PB) and petal fall (PF)

System Component	Time of year: Winter	M	A	M	J	J	A	S/O
Tree	Wood/buds		Wood/leaf/buds blossom BB GC PB PF		Wood/leaf/buds fruitlet		Wood/leaf/buds fruit	Wood/ buds
Codling/tortrix	Larva *bark*			Pupa	Adult/egg	Larva *fruit*	Adult/egg	Larva *fruit*
Winter moth	Adult/egg *bark*			Larva *buds, leaves blossoms*		Pupa *soil*		Adult
Sawfly	Larva *soil*	Pupa *soil*		Adult/egg *blossom*	Larva *fruitlet* *soil*			
Rosy apple aphid	Egg *bark*		Nymph/adult (2 gens) *leaves*				*On plantain*	Adult/egg *bark*
Woolly aphid	Nymph *bark*			Adult *bark*		Nymph/adult (3 gens) *leaves*		Nymph *bark*
Red spider mite	*Egg* *bark*					*Juvenile stages/adult/egg* (5 gens) *leaves*		*Egg* bark

Mildew	Mycelium *buds*	Mycelium/conidium (many gens) *blossoms, leaves* *buds, leaves*	Mycelium *buds*
Scab	Perithecium *leaves in soil*	Ascospore *aerial* / Mycelium/conidium (many gens) *blossoms/leaves, fruitlet/leaves, fruit/leaves*	Perithecium *leaves in soil*
Phytophthora	Mycelium/canker *soil/wood*	Mycelium/conidium/canker *soil/wood* *soil, fruit/wood*	Canker *wood*
Nectria	Canker/ascospore *wood*	Conidia/canker *wood* Conidia/canker *fruit/wood*	Canker *wood*
Gloeosporium	Canker *wood/leaves*	Canker *wood* Mycelium/canker *leaves/wood fruit/wood*	Canker *wood/leaves*
Fireblight	Canker *wood*	Bacterial growth/canker *blossoms, leaves/wood* *leaves/wood*	Canker *wood*
Phytoseiid mites	Adult ♀ *bark*	Egg/juvenile stages/adult (4 gens) *tree*	Adult ♀ *bark*
Anthocorids	Adults *bark*	Egg/nymph/adult (2 gens) *tree*	Adults *bark*
Mirids	Egg *bark*	Nymph *tree* Adult *tree*	Egg *bark*

Table 8.3 Damage matrix: the effect of pest components on the tree and fruit. Each * indicates an interaction which is detrimental

Tree components	**Codling/ Tortrix**	**Winter moth**	**Sawfly**	**Rosy apple aphid**	**Woolly aphid**	**Mites**	**Mildew**	**Scab**	**Nectria**	**Phytophora**	**Gloeo-sporium**	**Fireblight**
Vegetative buds				*			*					
Leaves				*	*	*	*	*				
Wood									*	*	*	*
Fruit buds		*		*			*					
Blossom trusses	*	*		*	*	*	*					
Fruitlet	*		*					*				
Fruit	*							*	*	*	*	
Fruit in storage								*	*	*	*	

Table 8.4 Pesticide impact matrix: the effect of pesticides on components of the orchard system. An * indicates negative effects on the components (mortality, reduced fungal development or damage to tree). An R indicates that resistant populations exist.

System Components	**Acaricides**		**Aphicides**	**Insecticides**		**Fungicides**		**Biological agents**
	Old (Kelthane)	**New (Tedion)**	**(Aphox)**	**Broad spect. (OP's)**	**Selective (Dimilin)**	**Old (Morocide)**	**New (Nimrod)**	**(Bacillus thuringiensis)**
Pests: Aphids			*	*				
Red spider Mite	*	*		*(R)		*		
Moths and Sawfly				*	*			*
Predators: Anthocorids and Mirids				*				
Phytoseiids	*			* (R)		*		
Diseases: Mildew						*	*	
Tree:							*	

Table 8.5 Effects of pesticides as orchard sprays, on *Typhlodromus pyri*. (From Cranham and Solomon 1981)

Pest/disease to be controlled	**Harmless *or* of low toxicity to typhs**	**Harmful to typhs**
Aphids (including woolly aphid)	*pirimicarb* (Aphox) demeton-S-methyl	vamidothion (Kilval)
Apple sucker	HCH; azinphos-methyl with demeton-S-methyl sulphone (Gusathion MS)	
Apple sawfly	HCH	
Winter moth Codling moth Fruit tree tortrix (*Archips*) Fruitlet mining tortrix (*Pammene*) Summer fruit tortrix (*Adoxophyes*)	*diflubenzuron* (Dimilin) (not for *Adoxophyes* control) carbaryl phosalone (Zolone) azinphos-methyl (Gusathion, Gusathion M5)	permethrin (Ambush) cypermethrin *deltamethrin (Decis)*
Spider mite	*tetradifon* (Tedion) *propargite* (Omite) cyhexatin (Plictran)	dicofol (Kelthane) amitraz (Mitac)
Apple rust mite	carbaryl	pirimiphos-methyl (Blex)
Mildew	*bupirimate* (Nimrod) *triadimefon* (Bayleton) *fenarimol* (Rubigan) thiophanate-methyl (Mildothane)	dinocap binapacryl (Morocide) Sulphur, wettable benomyl (Benlate)
Scab	*captan* *dodine* (Melprex) *dithianon (Delan-Col)* *urea* (in autumn)	mancozeb (Karamate) propineb (Antracol) benomyl (Benlate) carbendazim (Bavistin)
Storage rots	*captan*	
Fruit thinners	carbaryl *ethephon* (Ethrel)	
Branching agent	*MB25–105*	

Notes: When there is only one trade name for a given pesticide this is given in parenthesis.
The common chemical names of pesticides so far shown to be virtually harmless to *T. pyri* are in italics.

checked by density-dependent mortality factors at some stage in their life cycle. When a pesticide is applied before this critical stage, it may simply result in a proportionate decrease in this density-dependent mortality. Denholm (1981) suggests that as these regulatory factors operate primarily on the immature stages, control should be aimed at the adult stage rather than by applying ovicides or larvicides. Such a technique may be satisfactory against mosquitoes and tsetse flies, for example, where the adults are in a

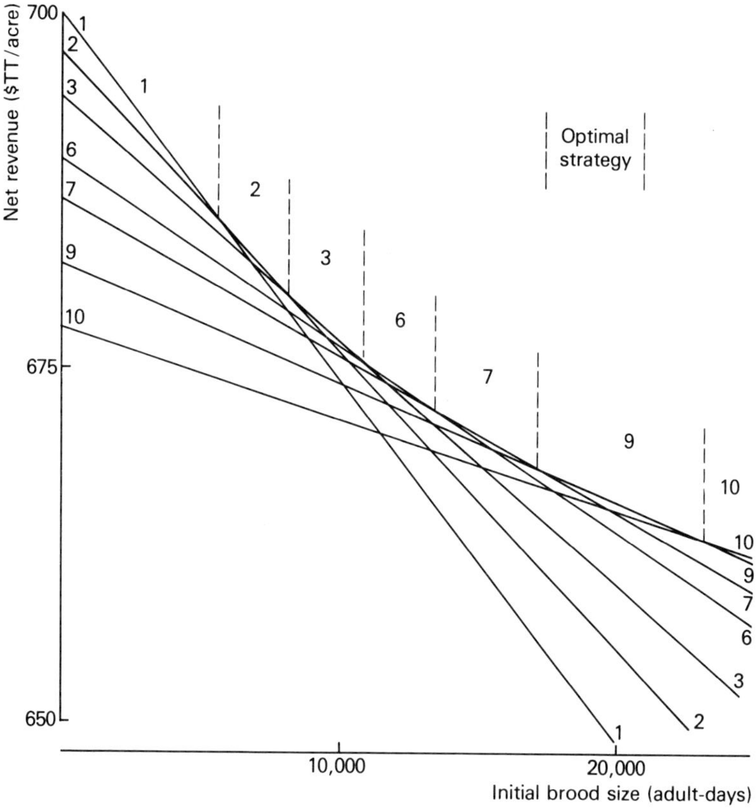

Fig. 8.2 Optimum spraying strategies against a single froghopper brood. (From Conway *et al.* 1975)

different habitat to the larvae stage, but is more difficult when protecting crops in which all stages of a pest may be present simultaneously. Denholm *et al.* (1983) report control of houseflies with bioresmethrin. Where possible a reduction of immigrants or control of adults before oviposition is needed but if adults are treated after mating, an opportunity exists for susceptible populations to mate with resistant individuals and dilute the level of resistance in the population which survives treatment.

Conservation of susceptible individuals is an important means of delaying resistance. Georghiou and Taylor (1977) refer to 'refugia' areas which escape treatment. These may be localized to the undersides of leaves within a treated area, or areas of crop which are deliberately left untreated. Such areas are sometimes left in perennial crops, such as citrus, to conserve natural enemies, but in most crops it is uneconomic to leave a part exposed to pest damage when inspection has clearly shown that treatment is necessary.

The risk of resistance to a pesticide is increased when a pest is exposed to the same type of pesticide on all host plants. Thus, *Heliothis* spp.

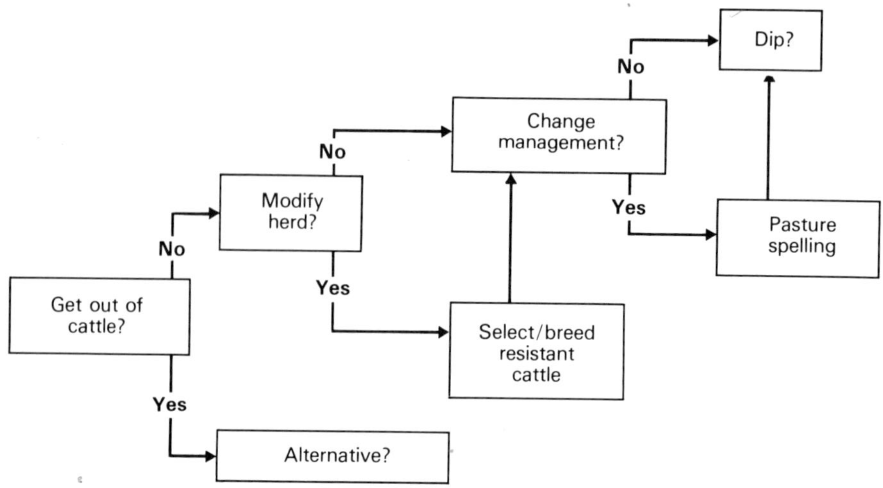

(a)

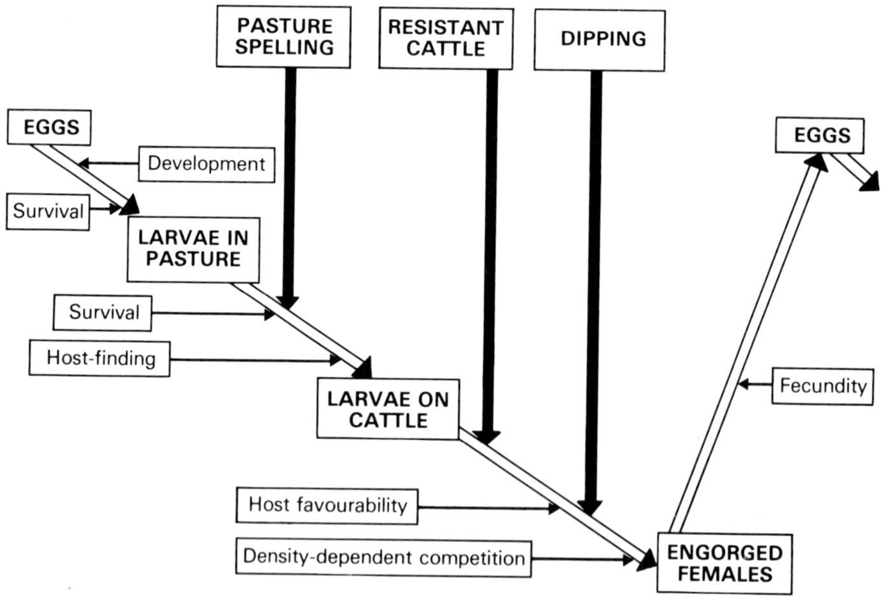

(b)

Fig. 8.3 (a) A simplified decision tree for tick control
(b) A simplified representation of the tick population model (From Norton, Sutherst and Maywald 1982)

populations resistant to insecticides have been found mostly in areas where not only cotton is treated, but also other crops such as maize, sorghum, tobacco and tomatoes. Development of resistance is delayed when susceptible individuals from untreated alternative host plants immigrate into the treated areas and mate with those which are resistant (Comins 1976).

Improved timing of pesticide treatments should also enable lower dosages than normally recommended to be applied. Graham-Bryce (1977) referred to the overdosing of populations if each droplet contained a lethal dose (see p. 116 Chapter 6). This suggests that where lower volumes of spray are applied in an optimum droplet-size range, the dosage should be spread over the average number of droplets that are likely to impact on a target. A smaller dosage will shorten the persistence of the pesticide in the environment, and reduce selection for resistance. Wolfe (1981) suggested that when a systemic fungicide is applied to seed, if might be effective to treat half the seed with the full rate of fungicide and mix this seed with untreated seed. He extended this idea by treating one variety with a fungicide and mixing this with the untreated seed of other varieties. The variety which is more disease-susceptible than others in a variety mixture should be treated. This is an interesting example of integrating chemical control with host-plant resistance which keeps costs of treatment to a minimum and increases crop diversity.

Resistance to a fungicide can occur very rapidly, e.g. in Europe matalaxyl was marketed in 1978 for control of late blight (*Phytophthora infestants*) on potatoes but by 1980 (Davidse *et al.* 1981) disease control failed in Holland, so other strategies being considered include the use of mixtures of fungicides with different modes of action or their use in rotation (Delp 1981). A low dosage of cypermethrin enhances the activity of the organophosphate diazinon (Dittrich 1981). This potentiation between insecticides may be exploited with other mixtures. Use of negatively correlated cross-resistance to a synthetic pyrethroid in organophosphorus-resistant pests has also been reported in mites (Chapman and Peerman 1979) and rice leafhoppers (Ozaki 1980). Some chemicals act as synergists by inhibition of known resistance mechanisms and this approach is regarded as a most promising future strategy for chemists to exploit (Dittrich 1981), but will require extensive evaluation as the effect of insecticidal mixtures is usually unpredictable (Sawicki 1981). In Japan, resistance to many organophosphorus insecticides in the green rice leafhopper *Nephottetix cincticeps* has been overcome by simultaneous application of the systemic organophosphorus fungicide kitazin which inhibits the esterases involved in OP (organophosphorus) resistance while providing control of the disease rice blast (Miyala *et al.* 1979).

Resistance to herbicides has not yet been a major problem, although resistance to triazine herbicides, mainly atrazine, has been reported since 1968 particularly in *Chenopodium album* and several *Amaranthus* spp., and in the UK resistance of *Senecio vulgaris* to simazine has been confirmed. Nevertheless, efficient use of one herbicide has led to serious new weed problems, thus control of broad-leaved weeds in cereals has increased the importance of grass weeds. In view of this more attention is being given to a weed management approach in which there is a rational deployment of physical and chemical methods to control weeds in a particular crop, in conjunction with appropriate cultural and husbandry practices to achieve control in a cropping system over a period of years (Fryer 1981). In the control of wild oats, a programme of herbicide treatments during the season was significantly better than a single application in terms of weed control, seed return and improved crop yield (Wilson and Cussans 1978). When the weed population has been effectively reduced, overall treatment may not be economic, but survival of some weeds increases the number of seeds in the soil for future infestations. Localized treatment of low populations of weeds is thus an important part of a weed management programme, hence the interest in equipment such as the 'rope-wick' applicator. Further research is clearly needed to examine the effects of different cropping systems and control strategies on changes in population of individual weed species.

More research is also needed on novel approaches to controlling pest populations. So far, many of the newer techniques referred to in Chapter 3 have been examined in isolation, but their most important contribution to pest management is more likely to be as just one component of an integrated control strategy. The newer techniques are not easy to examine under field conditions in small plots, so the more traditional means of evaluating the efficacy of a chemical described earlier are not always relevant. In many cases, such as the cattle tick example cited earlier, the use of systems analysis is the only method of evaluating different control strategies, because it would be impossible to carry out trials with all the possible strategies. Large-scale trials are very difficult to organize, very expensive, and require careful analysis. On cotton in Arkansas an insect management programme was established over 130 km^2 and this has expanded to include control of weeds and plant diseases, production and marketing technologies (Phillips 1978). These will range from studying predator or parasitoid/host interactions to improving delivery systems of pesticides, including microbials and pheromones. In addition to the large-scale approach, more detailed information on individual components of the system are required.

Decision-making by Government

Certain pest management strategies have been advocated by government for decades. The sowing and uprooting of cotton and the destruction of crop residues by specified dates has been a cultural control to check pink bollworm populations. Such a control is only effective on a regional or national scale. The issue of seed has also been regulated by governments for some crops in order to control which varieties are sown and allow seed treatment to be effectively organized. Treatment of cotton seed has been widely adopted to control bacterial blight (*Xanthomonas malvacearum*) disease.

Regulations control the movement of plant material in an effort to stop the introduction of new pests. Colorado beetle is a major pest of potatoes in continental Europe, which has so far not become established in the UK. Frequently, beetles are reported on cargoes in UK ports and occasionally infest fields, so these areas are rapidly treated with insecticides to eradicate it. Quarantine regulations are particularly important on geographical islands and also in North America, as a number of their major pests were imported from Europe.

The use of pesticides is regulated by government principally to protect the operators applying them, and the consumer. Instructions on the label should stipulate appropriate protective clothing and recommend the dosage rates and the interval between the last application and harvest, to ensure residues are below the specified level. Few government recommendations go further to limit the choice of chemical for particular pests, except where special restrictions limit the range of chemicals which can be applied aerially, so it is left to commercial companies to seek approval and registration of their products for the individual crop or pest. The user is sometimes faced with a long list of products without any reliable indication of which would be appropriate for use in an IPM programme such as the list shown in Table 8.5 (p. 176). In some countries, more restricted lists of pesticides have been recommended, e.g. against cotton pests in Central Africa, where a scheme to rotate particular acaricides around regions was adopted to try and delay the impact of resistance (Duncombe 1973). Furthermore, synthetic pyrethroids are not registered for winter-sown crops and their use on cotton is limited to a maximum period of nine weeks in any one season (Gledhill 1982). In Denmark, registration of a pesticide can be withdrawn if it becomes ineffective through resistance, and experimental data on cross-resistance and the potential for developing resistance is needed before registration of new products (Anon. 1980). Greater formal regulation of pesticides in this way may be needed to avoid multiple selection of resistance which would reduce the effective life of all the available pesticides.

Clearly, governments have a major role to play in developing improved methods of pest control by implementing policies which will encourage more ecologically orientated pest management programmes. Initially this requires an improvement in existing information, e.g. to advise use of less toxic and more selective insecticides where possible. Why use a toxic broad-spectrum organophosphate against aphids if a less hazardous and more selective product is available? Funding for research on all aspects of IPM should have priority, particularly as adequate long-term studies are needed to assess effects of different strategies used on a regional scale. Alongside research, more training will be required at all levels. Fryer (1981) pointed out that constraints on the adoption of weed management practice include lack of support for weed science as a scientific discipline. Clearly, if weed science is to be developed, education will have to play a vital role. Marketing will change from merely selling a product to providing a pest management service either by private consultants, or by technical specialists in the agrochemical industry. In some cases advice will come from specialists employed by the crop processors as in the case of sugar beet in the UK.

Pest management will not be easy. It will require more skills to replace the alternatives of doing-nothing-and-hope-for-the-best or opening a pesticide can, mixing the contents with water and applying it inefficiently, even if effectively, against a pest. Society is concerned about the lack of food for a large proportion of the world's population, also about current pesticide usage and its effects on the environment. The manipulation of different control techniques in pest management programmes will no doubt increase the sensible use of pesticides in some areas, particularly in the developing countries, to raise food production. In contrast, further expansion of IPM programmes in the USA, Europe, Japan and other countries using pesticides will result in their more efficient use. The real world is a dynamic place, so pest management will have to evolve to respond to the infinite challenges that pests will continue to offer man.

Appendix 1

Calculation of the regression line relating to probits and log-dose

The procedure for computing the regression line and LD_{50} using probit analysis (Finney 1952) is as follows:

1. Tabulate the dose applied (or concentration) in column 1 as shown in Table A1.1.
2. Column 2 gives the number of insects (or other pests) used in each test.
3. The number that died (or other response measured) is given in column 3.
4. The percentage mortality is calculated (col. 4) and these values are corrected in relation to the response in the control treatment by Abbott's formula. The corrected mortality is recorded in column 5.
5. The doses in column 1 are converted to logarithms (x). Negative values are avoided by adding one or more units to the logarithm as in column 6.
6. Using probit tables (see Table A1.3) empirical probit values corresponding to the percentages in column 5 are recorded in column 7.
7. The empirical probits (col. 7) are plotted against log-doses (col. 6) on ordinary graph paper and a provisional line is fitted to the points (see Fig. A1.1).
8. From this provisional line, expected probits Y are read for each value of x and tabulated in column 8.
9. The working probit y is calculated using the formula

$$y = y_0 + kp$$

   where y_0 and k are from Table A1.2 for the appropriate value of Y (col. 8), and p is the percentage mortality (col. 5). The values of y are recorded in column 9.
10. Weighting coefficients for each point Y are also read from Table A1.2 for the appropriate level of natural mortality and given in column 10. Each coefficient is multiplied by the number of insects used (col. 2) and the products (w) are recorded in column 11.

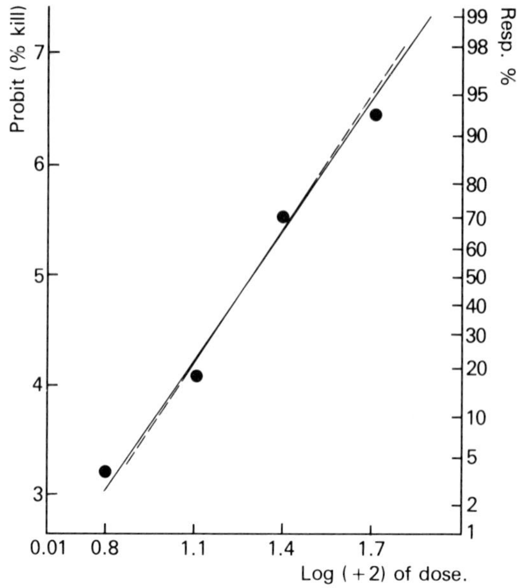

Fig. A1.1

11. Each of the values of w are multiplied first by the log-dose (col. 6) and recorded in column 12; then by the working probit y from column 9 and recorded in column 13.
12. The values in column 11 are added to give Σw, similarly addition of column 12 gives Σwx and column 13, Σwy. The means $\overline{x}$ and $\overline{y}$ are calculated:

$$x = \frac{\Sigma wx}{\Sigma w}; \overline{y} = \frac{\Sigma wy}{\Sigma w}$$

13. Σwx^2 is then calculated by multiplying wx by x on each line and summing the products.
14. In a similar way, calculate Σwy^2.
15. On each line multiply wx by y and sum the products to give Σwxy.
16. The slope of the regression line b can be calculated from

$$b = \frac{\Sigma wxy - x\Sigma wy}{\Sigma wx^2 - x\Sigma wx}$$

The regression equation is:

$$y = \overline{y} + b(x - \overline{x})$$

or $$y = (\overline{y} - b\overline{x}) + bx$$

17. Calculate values of y from this equation for each value of x and tabulate them in column 14. If the values in columns 14 and 7 (Y) do not differ by more than 0.2, the fit of the regression equation is adequate, otherwise the calculations will need to be repeated from 9. above.
18. The LD_{50} value is determined by calculating x where $y = 5.0$.

Table A1.1 Calculation of log-dose/probit regression line

1	2	3	4	5	6	7	8	9	10	11	12	13	14
Dose or % concentration	No. of test insects	No. of insects dead	% response (death)	Corrected mortality %	log (+2) of dose	Empirical probit	Expected probit	Working probit	Weighting coefficient	Weight			Calculated values from the regression line
	n				x		Y	y		w	wx	wy	Y^1
0.5	47	44	93.6	93	1.7	6.48	6.65	6.46	0.218	10.2	17.34	65.9	6.57
0.25	50	36	72	70	1.4	5.52	5.40	5.52	0.530	26.5	37.1	146.28	5.41
0.125	54	13	24.1	18	1.1	4.08	4.30	4.11	0.320	17.3	19.0	71.10	4.24
0.0625	34	4	11.7	4	0.8	3.25	3.05	3.27	0.038	1.3	1.04	4.25	3.07
0	50	4	8.0										

$\Sigma w = 55.3$ $\Sigma wx = 74.48$ $\bar{x} = 1.3468$ $\Sigma wy = 287.52$ $\bar{y} = 5.199$
$\Sigma wx^2 = 103.15$ $\Sigma wy^2 = 1539.298$ $\Sigma wxy = 398.299$
$b = 3.893$
Regression equation $y = 3.8928x - 0.0438$
If $y = 5.0$ then $x = 1.296$ (this corresponds to a dose of 0.135)
Variance = 0.00125; $\chi^2 = 1.398$ (with 2 deg. of freedom, P 0.5)

Table A1.2 Factors for calculating working probits and weighting coefficients. (After Swaroop and Uemura)

Expected probit	Factors for working probit		Weighting coefficient for different levels of natural mortality		
Y	y_0	k	0.00	0.08	0.15
1.6	1.33	8.115	0.005		
1.7	1.42	5.805	0.006		
1.8	1.51	4.194	0.008		
1.9	1.60	3.061	0.011		
2.0	1.70	2.256	0.015		
2.1	1.79	1.6800	0.019		
2.2	1.88	1.2634	0.025	0.001	
2.3	1.97	0.9596	0.031	0.001	0.001
2.4	2.06	0.7362	0.040	0.002	0.001
2.5	2.15	0.5705	0.050	0.003	0.002
2.6	2.23	0.4465	0.062	0.005	0.003
2.7	2.32	0.3530	0.076	0.008	0.004
2.8	2.41	0.2819	0.092	0.013	0.007
2.9	2.49	0.2274	0.110	0.19	0.010
3.0	2.58	0.1852	0.131	0.027	0.015
3.1	2.66	0.1524	0.154	0.038	0.022
3.2	2.74	0.1267	0.180	0.053	0.030
3.3	2.83	0.1063	0.208	0.070	0.042
3.4	2.91	0.0902	0.238	0.092	0.056
3.5	2.98	0.0772	0.269	0.117	0.074
3.6	3.06	0.0668	0.302	0.145	0.095
3.7	3.14	0.0584	0.336	0.177	0.119
3.8	3.21	0.0515	0.370	0.211	0.146
3.9	3.28	0.0459	0.405	0.247	0.176
4.0	3.34	0.0413	0.439	0.283	0.208
4.1	3.41	0.0376	0.471	0.320	0.241
4.2	3.47	0.0345	0.503	0.356	0.274
4.3	3.53	0.0320	0.532	0.391	0.307
4.4	3.58	0.0300	0.558	0.424	0.339
4.5	3.62	0.0284	0.581	0.453	0.370
4.6	3.66	0.0272	0.601	0.480	0.397
4.7	3.70	0.0262	0.616	0.502	0.421
4.8	3.72	0.0256	0.627	0.520	0.442
4.9	3.74	0.0252	0.634	0.534	0.458
5.0	3.75	0.0251	0.637	0.542	0.471

The working probit y is computed according to the formula $y = y_0 + kp$, where p is the percentage mortality.

Notes

1. y_0 is from Fisher and Yates (1974), Table IX_2, column 2, p. 71.
2. Weighting coefficients are from Fisher and Yates (1974), table IX_3, p. 72–3 which has values for other leaves of natural mortality.
3. Swaroop, S. and Uemura, K. (1956) Wld Hlth Org. Mal/178 Mimeog.

Table A1.2 (continued)

Expected probit	Factors for working probit		Weighting coefficient for different levels of natural mortality		
Y	y_0	k	0.00	0.08	0.15
5.0	3.75	0.0251	0.637	0.542	0.471
5.1	3.74	0.0252	0.634	0.546	0.478
5.2	3.72	0.0256	0.627	0.546	0.481
5.3	3.68	0.0262	0.616	0.540	0.479
5.4	3.62	0.0272	0.601	0.530	0.473
5.5	3.54	0.0284	0.581	0.516	0.463
5.6	3.42	0.0300	0.558	0.498	0.449
5.7	3.27	0.0320	0.532	0.477	0.431
5.8	3.08	0.0345	0.503	0.453	0.411
5.9	2.83	0.0376	0.471	0.426	0.388
6.0	2.52	0.0413	0.439	0.398	0.363
6.1	2.13	0.0459	0.405	0.368	0.336
6.2	1.64	0.0515	0.370	0.337	0.309
6.3	1.03	0.0584	0.336	0.306	0.281
6.4	0.26	0.0668	0.302	0.276	0.253
6.5	−0.71	0.0772	0.269	0.246	0.226
6.6	−1.92	0.0902	0.238	0.218	0.200
6.7	−3.46	0.1063	0.208	0.190	0.175
6.8	−5.41	0.1267	0.180	0.165	0.152
6.9	−7.90	0.1524	0.154	0.142	0.131
7.0	−11.10	0.1852	0.131	0.120	0.111
7.1	−15.23	0.2274	0.110	0.101	0.093
7.2	−20.60	0.2819	0.092	0.084	0.078
7.3	−27.62	0.3530	0.076	0.070	0.064
7.4	−36.89	0.4465	0.062	0.057	0.052
7.5	−49.20	0.5705	0.050	0.046	0.042
7.6	−65.68	0.7362	0.040	0.037	0.034
7.7	−87.93	0.9596	0.031	0.029	0.027
7.8	−118.22	1.2634	0.025	0.023	0.021
7.9	−159.79	1.6800	0.019	0.018	0.016
8.0	−217.3	2.256	0.015	0.013	0.012
8.1	−297.7	3.061	0.011	0.010	0.009
8.2	−410.9	4.194	0.008	0.008	0.007
8.3	−571.9	5.805	0.006	0.006	0.005
8.4	−802.8	8.115	0.005	0.004	0.004

19. A regression line can be drawn if some other value of x is determined, e.g. when y = 7.0.
20. The variance V of m, the estimated value of log LD_{50}, is calculated from the equation:

$$V = \frac{1}{b^2}\left(\frac{1}{\Sigma w} \frac{(m - \bar{x})^2}{\Sigma wx^2 - \frac{(\Sigma wx)^2}{\Sigma w}}\right)$$

21. The homogeneity of the data is tested by calculating χ^2 from:

$(\Sigma wy^2 - y\Sigma wy) - b\ (\Sigma wxy - x\Sigma wy)$

The value of χ^2 is compared with tables of this statistic for n – 2 deg. of freedom where n is the number of doses tested.
Heterogeneity is indicated if the calculated value of χ^2 is greater than the value for the 5 per cent level in the tables.
22. The fiducial limits m_1, m_2, with 95 per cent confidence are calculated from the variance (V) using the formulae

$m_1 = m - 1.96\sqrt{V}$
$m_2 = m + 1.96\sqrt{V}$

These limits are acceptable if g (see Finney 1952) is sufficiently small.

Table A1.3 The probit transformation

% kill	0	1	2	3	4	5	6	7	8	9
00	—	2.67	2.95	3.12	3.25	3.36	3.45	3.52	3.59	3.66
10	3.72	3.77	3.82	3.87	3.92	3.96	4.01	4.05	4.08	4.12
20	4.16	4.19	4.23	4.26	4.29	4.33	4.36	4.39	4.42	4.45
30	4.48	4.50	4.53	4.56	4.59	4.61	4.64	4.67	4.69	4.72
40	4.75	4.77	4.80	4.82	4.85	4.87	4.90	4.92	4.95	4.97
50	5.00	5.03	5.05	5.08	5.10	5.13	5.15	5.18	5.20	5.23
60	5.25	5.28	5.31	5.33	5.36	5.39	5.41	5.44	5.47	5.50
70	5.52	5.55	5.58	5.61	5.64	5.67	5.71	5.74	5.77	5.81
80	5.84	5.88	5.92	5.95	5.99	6.04	6.08	6.13	6.18	6.23
90	6.28	6.34	6.41	6.48	6.55	6.64	6.75	6.88	7.05	7.33

(Abbreviated version of Table IX in Fisher and Yates, 1974, Pages 68–70).

Appendix 2

Random numbers

Table A2.1 Random numbers

03 47 43 73 86	36 96 47 36 61	46 98 63 71 62	33 26 16 80 45	60 11 14 10 95
97 74 24 67 62	42 81 14 57 20	42 53 32 37 32	27 07 36 07 51	24 51 79 89 73
16 76 62 27 66	56 50 26 71 07	32 90 79 78 53	13 55 38 58 59	88 97 54 14 10
12 56 85 99 26	96 96 68 27 31	05 03 72 93 15	57 12 10 14 21	88 26 49 81 76
55 59 56 35 64	38 54 82 46 22	31 62 43 09 90	06 18 44 32 53	23 83 01 30 30
16 22 77 94 39	49 54 43 54 82	17 37 93 23 78	87 35 20 96 43	84 26 34 91 64
84 42 17 53 31	57 24 55 06 88	77 04 74 47 67	21 76 33 50 25	83 92 12 06 76
63 01 63 78 59	16 95 55 67 19	98 10 50 71 75	12 86 73 58 07	44 39 52 38 79
33 21 12 34 29	78 64 56 07 82	52 42 07 44 38	15 51 00 13 42	99 66 02 79 54
57 60 86 32 44	09 47 27 96 54	49 17 46 09 62	90 52 84 77 27	08 02 73 43 28
18 18 07 92 46	44 17 16 58 09	79 83 86 19 62	06 76 50 03 10	55 23 64 05 05
26 62 38 97 75	84 16 07 44 99	83 11 46 32 24	20 14 85 88 45	10 93 72 88 71
23 42 40 64 74	82 97 77 77 81	07 45 32 14 08	32 98 94 07 72	93 85 79 10 75
52 36 28 19 95	50 92 26 11 97	00 56 76 31 38	80 22 02 53 53	86 60 42 04 53
37 85 94 35 12	83 39 50 08 30	42 34 07 96 88	54 42 06 87 98	35 85 29 48 39
70 29 17 12 13	40 33 20 38 26	13 89 51 03 74	17 76 37 13 04	07 74 21 19 30
56 62 18 37 35	96 83 50 87 75	97 12 25 93 47	70 33 24 03 54	97 77 46 44 80
99 49 57 22 77	88 42 95 45 72	16 64 36 16 00	04 43 18 66 79	94 77 24 21 90
16 08 15 04 72	33 27 14 34 09	45 59 34 68 49	12 72 07 34 45	99 27 72 95 14
31 16 93 32 43	50 27 89 87 19	20 15 37 00 49	52 85 66 60 44	38 68 88 11 80
67 34 30 13 70	55 74 30 77 40	44 22 78 84 26	04 33 46 09 52	68 07 97 06 57
74 57 25 65 76	59 29 97 68 60	71 91 38 67 54	13 58 18 24 76	15 54 55 95 52
27 42 37 86 53	48 55 90 65 72	96 57 69 36 10	96 46 92 42 45	97 60 49 04 91
00 39 68 29 61	66 37 32 20 30	77 84 57 03 29	10 45 65 04 26	11 04 96 67 24
29 94 98 94 24	68 49 69 10 82	53 75 91 93 30	34 25 20 57 27	40 48 73 51 92
16 90 82 66 59	83 62 64 11 12	67 19 00 71 74	60 47 21 29 68	02 02 37 03 31
11 27 94 75 06	06 09 19 74 66	02 94 37 34 02	76 70 90 30 86	38 45 94 30 38
35 24 10 16 20	33 32 51 26 38	79 78 45 04 91	16 92 53 56 16	02 75 50 95 98
38 23 16 86 38	42 38 97 01 50	87 75 66 81 41	40 01 74 91 62	48 51 84 08 32
31 96 25 91 47	96 44 33 49 13	34 86 82 53 91	00 52 43 48 85	27 55 26 89 62

56 67 40 67 14	64 05 71 95 86	11 05 65 09 68	76 83 20 37 90	57 16 00 11 66
14 90 84 45 11	75 73 88 05 90	52 27 41 14 86	22 98 12 22 08	07 52 74 95 80
68 05 51 18 00	33 96 02 75 19	07 60 62 93 55	59 33 82 43 90	49 37 38 44 59
20 46 78 73 90	97 51 40 14 02	04 02 33 31 08	39 54 16 49 36	47 95 93 13 30
64 19 58 97 79	15 06 15 93 20	01 90 10 75 06	40 78 78 89 62	02 67 74 17 33
05 26 93 70 60	22 35 85 15 13	92 03 51 59 77	59 56 78 06 83	52 91 05 70 74
07 97 10 88 23	09 98 42 99 64	61 71 62 99 15	06 51 29 16 93	58 05 77 09 51
68 71 86 85 85	54 87 66 47 54	73 32 08 11 12	44 95 92 63 16	29 56 24 29 48
26 99 61 65 53	58 37 78 80 70	42 10 50 67 42	32 17 55 85 74	94 44 67 16 94
14 65 52 68 75	87 59 36 22 41	26 78 63 06 55	13 08 27 01 50	15 29 39 39 43
17 53 77 58 71	71 41 61 50 72	12 41 94 96 26	44 95 27 36 99	02 96 74 30 83
90 26 59 21 19	23 52 23 33 12	96 93 02 18 39	07 02 18 36 07	25 99 32 70 23
41 23 52 55 99	31 04 49 69 96	10 47 48 45 88	13 41 43 89 20	97 17 14 49 17
60 20 50 81 69	31 99 73 68 68	35 81 33 03 76	24 30 12 48 60	18 99 10 72 34
91 25 38 05 90	94 58 28 41 36	45 37 59 03 09	90 35 57 29 12	82 62 54 65 60
34 50 57 74 37	98 80 33 00 91	09 77 93 19 82	74 94 80 04 04	45 07 31 66 49
85 22 04 39 43	73 81 53 94 79	33 62 46 86 28	08 31 54 46 31	53 94 13 38 47
09 79 13 77 48	73 82 97 22 21	05 03 27 24 83	72 89 44 05 60	35 80 39 94 88
88 75 80 18 14	22 95 75 42 49	39 32 82 22 49	02 48 07 70 37	16 04 61 67 87
90 96 23 70 00	39 00 03 06 90	55 85 78 38 36	94 37 30 69 32	90 89 00 76 33

(From Fisher and Yates, 1974)

Appendix 3

Conversion tables

Table A3.1

	A	B	A→B	B→A
Weight	oz	g	× 28.35	× 0.0353
	lb	kg	× 0.454	× 2.205
	cwt	kg	× 50.8	× 0.0197
	ton (long)	kg	× 1016	× 0.000984
	ton (short)	ton (long)	× 0.893	× 1.12
Surface area	in^2	cm^2	× 6.45	× 0.155
	ft^2	m^2	× 0.093	× 10.764
	yd^2	m^2	× 0.836	× 1.196
	yd^2	acres	× 0.000207	× 4840
	acres	ha	× 0.405	× 2.471
Length	μm	mm	× 0.001	× 1000
	in	cm	× 2.54	× 0.394
	ft	m	× 0.305	× 3.281
	yd	m	× 0.914	× 1.094
	mile	km	× 1.609	× 0.621
Velocity	ft/s	m/s	× 0.305	× 3.281
	ft/min	m/s	× 0.00508	× 197.0
	mile/hr	km/hr	× 1.609	× 0.621
	mile/hr	ft/min	× 88.0	× 0.0113
	knot	ft/s	× 1.689	× 0.59
	m/s	km/hr	× 3.61	× 0.277
	cm/s	km/hr	× 0.036	× 27.78
Quantities/area	lb/acre	kg/ha	× 1.12	× 0.894
	lb/acre	mg/ft^2	× 10.4	× 0.09615
	kg/ha	mg/m^2	× 100	× 0.01
	mg/ft^2	mg/m^2	× 10.794	× 0.093
	oz/yd^2	cwt/acre	× 2.7	× 0.37

	A	B	A→B	B→A
	gal (Imp.)/acre	litres/ha	× 11.23	× 0.089
	gal (USA)/acre	litres/ha	× 9.346	× 0.107
	fl. oz (Imp.)/acre	ml/ha	× 70.05	× 0.0143
	fl. oz (USA)/acre	ml/ha	× 73.14	× 0.0137
	oz/acre	g/ha	× 70.05	× 0.0143
	oz/acre	kg/ha	× 0.07	× 14.27
Dilutions	fl. oz/100 gal (Imp.)	ml/100 litres	× 6.25	× 0.16
	pint/100 gal (Imp.)	ml/100 litres	× 125	× 0.008
	oz/gal (Imp.)	g/litre	× 6.24	× 0.16
	oz/gal (USA)	g/litre	× 7.49	× 0.134
	lb/100 gal (Imp.)	kg/100 litres	× 0.0998	× 10.02
Density of water	gal (Imp.)	lb	× 10	× 0.1
	gal (USA)	lb	× 8.32	× 0.12
	lb	ft^3	× 0.016	× 62.37
	litre	kg	× 1	× 1
	ml	g	× 1	× 1
	lb/gal (Imp.)	g/ml	× 0.0997	× 10.03
	lb/gal (USA)	g/ml	× 0.1198	× 8.34
	lb/ft^3	kg/m^3	× 16.1	× 0.0624
Volume	in^3	ft^3	× 0.000579	× 1728
	ft^3	yd^3	× 0.037	× 27
	yd^3	m^3	× 0.764	× 1.308
	fl. oz (Imp.)	ml	× 28.35	× 0.0352
	fl. oz (USA)	ml	× 29.6	× 0.0338
	gal (Imp.)	gal (USA)	× 1.20	× 0.833
	gal (Imp.)	litre	× 4.55	× 0.22
	gal (USA)	litre	× 3.785	× 0.264
	cm^3	m^3	$\times 10^{-6}$	$\times 10^{6}$
	cm^3	μm^3	$\times 10^{12}$	$\times 10^{-12}$
Pressure	lb/in^2	kg/cm^2	× 0.0703	× 14.22
	lb/in^2	bar	× 0.0689	× 14.504
	bar	kPa	× 100	× 0.01
	lb/in^2	kPa	× 6.89	× 0.145
	kN/m^2	kPa	× 1	× 1
	N/m^2	kPa	× 0.001	× 1000
	lb/in^2	atm	× 0.068	× 14.696
Power	hp	kW	× 0.7457	× 1.341
Temperature	°C	°F	$(\frac{9}{5}°C) + 32$	$\frac{5}{8}(°F - 32)$

(From Matthews 1969a)

Table A3.2 Local measurements
(Most countries use the metric system, but some local measurements are used.)

A		B	A → B	B → A
Manzana	(Argentine)	ha	× 1.00	× 1.00
Manzana	(Nicaragua)	ha	× 0.75	× 1.33
Manzana	(other countries)	ha	× 0.699	× 1.43
Fanegada	(Peru)	ha	× 2.89	× 0.346
Fanegada	(C. America)	ha	× 0.64	× 1.563
Feddan	(Sudan)	ha	× 0.42	× 2.38
Morgen	(S. Africa)	ha	× 0.856	× 1.168
Rai	(Thailand)	ha	× 0.16	× 6.25
Cuerda	(Venezuela)	m^2	× 7.14	× 0.14

Table A3.3 Characteristic parameters of spray droplets (volume, surface, number and terminal velocity) based on an application rate of 1 litre/ha on a flat surface

Monosized droplet diameter μm	*Droplet volume μm^3	No. of droplets n/cm^2	Droplet surface of 1 litre spray liquid (m^2)	Terminal velocity of water droplets in air (cm/s)
10	524	19,098.5	600.0	0.3
20	4,189	2,387.3	300.0	1.2
30	14,137	707.4	200.0	2.6
40	33,510	298.4	150.0	4.6
50	65,450	152.8	120.0	7.0
60	113,097	88.4	100.0	10.0
70	179,594	55.7	85.7	13.5
80	268,083	37.3	75.0	17.0
90	381,704	26.2	66.7	21.0
100	523,599	19.1	60.0	25.5
150	1,767,146	5.7	40.0	48.0
200	4,188,790	2.4	30.0	72.0
250	8,181,231	1.2	24.0	96.0
300	14,137,167	0.7	20.0	122.0

Total number of droplets for 1 litre/ha:

* 1 $\mu m^3 = 10^{-9}$ n/cm^2.100 million or n/$cm^2.10^8$

Appendix 4

Useful calculations

Pesticide calculations

1. To determine the quantity of a formulation (X) required to apply the recommended amount of active ingredient per hectare (A) with a formulation containing B percentage active ingredient.

$$\frac{A \times 100}{B} = X$$

Example: Apply 0.25 kg ai/ha of 5 per cent carbofuran granules

$$\therefore \quad \frac{0.25 \times 100}{5} = 5 \text{ kg granules/ha}$$

2. To determine the quantity of active ingredient (Y) required to mix with a known quantity of diluent (Q) to obtain a given concentration of spray

$$Q \times \frac{\text{per cent concentration required}}{\text{per cent concentration of active ingredient}} = Y$$

(*a*) *Example*: Mix 100 litres of 0.5 per cent ai, using a 50 per cent wettable powder

$$100 \times \frac{0.5}{50} = 1 \text{ kg of wettable powder}$$

(*b*) *Example*: Mix 2 litres of 5 per cent ai using a 75 per cent wettable powder

$$2000 \times \frac{5}{75} = 133 \text{ g of wettable powder}$$

Rate of application

The application rate will depend on
1. Number of nozzles on the spray bar or lance (N).
2. The output per nozzle in litres/min (Qt).
3. The width (swath) being sprayed (S).
4. The speed of the sprayer (V) in km/ha.

$$A = \frac{600 \times N \times Qt}{S \times V}$$

Note: 600 = 10,000 m^2/16.66 m/min (1 km/h)

Example

1. *To determine application rate*
   A sprayer has 20 nozzles applying 1.4 litres/min. If the swath is 12 m and the speed is 7 km/hr, what is the application rate per hectare?

$$\frac{600 \times 20 \times 1.4}{12 \times 7} = 200 \text{ litres/ha}$$

2. *To determine forward speed*
   If a sprayer has 20 nozzles each delivering 1.4 litres min and a swath of 12 m is treated, what speed has to be travelled to apply 200 litres/ha?

$$\frac{600 \times 20 \times 1.4}{12 \times 200} = 7 \text{ km/h}$$

Flow rate of nozzle in relation to pressure

Volume of spray liquid passing through a hydraulic nozzle is proportional to the square root of the pressure at the nozzle:

$$Q_1 = Q_2 \sqrt{\frac{P_1}{P_2}}$$

where Q_1 = flow rate (litres/min)
Q_2 = required flow rate
P_1 = pressure (bar or psi)
P_2 = required pressure

i.e. To double flow rate, pressure must be increased fourfold, but as pressure effects the droplet spectrum, a nozzle with different flow rate should be used unless the change in volume required is within 5 per cent.

Example
If output from a spray boom is 21 litres/min at 3 bar and required output is 20 litres/min, then the pressure can be changed to 2.8 bar.

References

Abbott, W. S. (1925) A method of computing the effectiveness of an insecticide, *J. Econ. Ent.*, **18**, 265–7.

Ables, J. R. and Ridgway, R. L. (1981) Augmentation of entomophagous arthropods to control pest insects and mites, in Papavizas, G. C. (ed.) *Biological Control in Crop Production*. Allanheld Osman, London, pp. 273–303.

Adams, J. B. and Drew, M. E. (1965) Grain aphids in New Brunswick III: aphid populations in herbicide-treated oat fields, *Can. J. of Zool.*, **43**, 789–94.

ADAS (1976) *The Utilization and Performance of Field Crop Sprayers 1976*. Farm Mechanization Studies No. 29. Ministry of Agriculture, Fisheries and Food (32 pp.).

Alford, D. V., Carden, P. W., Dennis, E. B., Gould, H. J. and Vernon, J. D. R. (1979) Monitoring codling and tortrix moths in United Kingdom apple orchards using pheromone traps, *Ann. Appl. Biol.*, **91**, 165–78.

Amsden, R. C. (1960) A laboratory apparatus for the continuous production of uniform droplets, *Agricultural and Veterinary Chemicals*, **1**, 139–40.

Amsden, R. C. (1970) The Metering and Dispensing of Granules and Liquid Concentrates. *BCPC Monograph* 2, pp. 124–9.

Anon (1980) Statutory order No. 410 of 17 September 1980 on pesticides. Ministry of the Environmental Protection, Translation LK, October (45 pp.).

Anon. (1981a) *Selected Environmental Law Statutes*. West Publishing Co. St Paul, Minnesota.

Anon. (1981b) *Pesticides* (5th edn), Council of Europe, Strasbourg.

Arnold, A. J. (1965) A high-speed automatic micrometer syringe, *J. Sci. Ins.* **42**, 350–51.

Arnold, A. J. (1967) Hand-operated micro-applicator to deliver drops of five sizes, *Lab. Practice*, **16**, 56–7.

Arnold, A. J., Needham, P. H. and Stevenson, J. H. (1973) A self-propelled portable insect suction sampler, *Ann. Appl. Biol.* **75**, 229–33.

Arnold, A. J. and Pye, B. J. (1980) Spray Application with Charged Rotary Atomizers. *BCPC Monograph* 24, pp. 109–17.

Arnold, A. J. and Pye, B. J. (1981) Electrostatic spraying of crops with the APE80, *Proc. 1981 Br. Crop Prot. Conf. – Pests and Diseases*, 2661–6.

Ashby, D. G. (1983) The principles of insecticide and fungicide evaluation, in N. Scopes and M. Ledieu (eds.), Pest and Disease Handbook, BCPC, pp. 53–103.

Asquith, D., Croft, B. A., Hoyt, S. C., Glass, E. H. and Rice, R. E. (1980) The systems approach and general accomplishments toward better insect control in pome and stone fruits, in: C. B. Huffaker (ed.), *New Technology of Pest Control*. Wiley Inter-Science, pp. 249–317.

Bailey, N. T. J. (1959) *Statistical Methods in Biology*. English Universities Press, London.

Bainbridge, A. and Jenkyn, J. F. (1976) Mildew reinfection in adjacent and separated plots of sprayed barley, *Ann. Appl. Biol.*, **82**, 477–84.

Baker, R. (1979) Recent developments in pheromone chemistry: commercial possibilities, *Proc. 1979 Br. Crop Prot. Conf. – Pests and Diseases*, **3**, 843–52.

Bals, E. J. (1975a) Development of a CDA herbicide handsprayer, *PANS* **21**,345–9.

Bals, E. J. (1975b) The importance of controlled droplet application (CDA) in pesticide applications, *Proc. 8th Br. Insectic. Fungic. Conf.*, 153–60.

Bals, E. J. (1978) Reduction of Active Ingredient Dosage by Selecting Appropriate Droplet Size for the Target. *BCPC Monograph* 22, 101–106.

Banks, C. J. and McCaulay, E. D. M. (1976) *Effects of Aphis fabae* and of its attendants ants and insect predators on yields on field beans (*Vicia faba* (L.)), *Ann. Appl. Biol.*, **60**, 445–53.

Bardner, R. (1978) Pest control in coffee, *Pestic. Sci 9* 458–64.

Bardner, R. and Fletcher, K. C. (1974) Insect infestations and their effects on the growth and yield of field crops: a review, *Bull. Ent. Res.*, **64**, 141–60.

Bardner, R., French, R. A., and Dupuch, M. J. (1981) *Agricultural Benefits of the Rothamsted Aphid Bulletin*, Rothamsted Report for 1980, Part 2.

Barnes, T. R. (1981) The use of chlortoluron for the control of *Alopecurus myosuroides, Proc. of Grass Weeds in Cereals in the U.K. Conf.*, 215–22.

Barrett, P. R. F. (1974) A spraying rig for the experimental application of herbicide to the floating leaves of water plants *Weed Res.*, **14**, 313–5.

Barrett, P. R. F. (1978) Some studies on the use of alginates for the placement and controlled release of diquat on submerged aquatic plants, *Pestic. Sci.*, **9** 425–33.

Barrett, P. R. F. and Logan, P. (1982) The localized control of submerged aquatic weeds in lakes with diquat-alginate, *Proc. EWRS 6th Symposium on Aquatic Weeds*, 193–8.

Barrett, P. R. F. and Murphy, K. J. (1982) The use of diquat-alginate for weed control in flowing waters, *Proc. EWRS 6th Symposium in Aquatic Weeds*, 200–8.

Bartlett, B. R. (1963) The contact toxicity of some pesticide residues to hymenopterous parasites and coccinellid predators, *J. Econ. Ent.*, **56**, 694–8.

Bartlett, J. T. and Bainbridge, A. (1978) Volumetric sampling of micro-organisms in the atmosphere, in: P. R. Scott and A. Bainbridge (eds), *Plant Disease Epidemiology*. Blackwell, pp. 23–30.

Baybee, T. A. and Greulach, L. (1980) The Mini-Hustler, *Proc. N.C. Weed Cont. Conf.*, 35.

Beaumont, A. (1947) The dependence on the weather of the dates of outbreaks of potato blight epidemics, *Trans. Br. Mycol. Sci.*, **31**, 45–53.

Bedford, G. O. (1980) Biology, ecology and control of palm rhinoceros beetles, *Ann. Rev. Ent.*, **25**, 309–39.

Bedford, G. O. (1981) Control of rhinoceros beetle by baculovirus, in: H. D. Burges, (ed.), *Microbial Control of Pests and Plant Diseases 1970–1980*. Academic Press, London and New York, pp. 409–26.

Beevor, P. S. and Campion, D. G. (1979) The field use of inhibitory components of lepidopterous sex pheromones and pheromone mimics, in: F. J. Ritter (ed.), *Chemical Ecology: odour communication in animals*. Elsevier North Holland Biomedical Press, Amsterdam, pp. 313–25.

Beglyarov. G. A. and Smetnik, I. (1977) Seasonal colonization of entomophages in the USSR, in: R. L. Ridgway and S. B. Vinson (eds), *Biological Control of Augmentation of Natural Enemies*. Plenum, New York, pp. 283–328.

Beingolea, G. O. D. (1977) Considerations on biological control and predation, *Revista Peruana de Entomologia*, **20**, 33–48.
Beroza, M., Paszek, E. C., Mitchell, E. R., Bierl, B. A., McLaughlin, J. R. and Chambers, D. L. (1974) Tests of a 3-layer laminated plastic bait dispenser for controlled emission of attractants from insect traps, *J. Env. Entomol.*, **3**, 926.
Berry, G., Cleaver, T. J., Nelder, J. A. and Salter, P. J. (1966) Methods of recording data in the laboratory and field, *Expl. Agric.*, **2**, 69–80.
Best, K. F. and Wenhardt, A. (1960) An experimental plot sprayer, *Weeds*, **8**, 464–5.
Blair, A. M. and Dean, M. L. (1976) Improvement in selectivity of Perfluidone against *Rottboellia exaltata* in maize with herbicide protectants, *Weed Res.*, **16**, 47–52.
Blair, A. M., Parker, C. and Kasasian, L. (1976) Herbicide protectants and antidotes – a review, *PANS*, **22**, 65–74.
Blair, A. M., Taylor, W. A. and Tottman, D. R. (1975) Development of a new small plot sprayer and some predications of the field performance, *Weed Res.*, **15**, 185–8.
Blanchard, A. D. (1979) Spraying six chemicals without reloading, *Farm Chemicals* (June), 40–1.
Bliss, C. I. (1935) The calculation of the dosage–mortality curve, *Ann. Appl. Biol.*, **22**, 134–67.
Boize, L. M. and Dombrowski, N. (1976) The atomization characteristics of a spinning disc ultra low volume applicator, *J. Agric. Engng. Res.*, **21**, 87–99.
Bourgeois, F. (1978) *Small Plot Spraying Machine*. Ciba Geigy Documents, AC 2.72.
Bouse, L. F. (1975) Use of pulsed jets to atomize various liquids, *Trans. Ann. Soc. Agric. Eng.*, **18**, 618–22.
Bowden, J. (1973a) The influence of moonlight on catches of insects in light-traps in Africa. Part 1: The moon and moonlight, *Bull. Ent. Res.*, **63**, 113–28.
Bowden, J. (1973b) The significance of moonlight in photoperiodic responses of insects, *Bull. Ent. Res.*, **62**, 605–12.
Bowden, J. and Church, B. M. (1973) The influence of moonlight on catches of insects in light-traps in Africa. Part 2: The effect of moon phase on light trap catches, *Bull. Ent. Res.*, **63**, 129–42.
Bowden, J. and Morris, M. G. (1975) The influence of moonlight on catches of insects in light-traps in Africa. Part 3: The effective radius of a mercury vapour light trap and the analysis of catches using effective radius, *Bull. Ent. Res.*, **65**, 303–48.
Bradfield, K. W. (1980) A self-propelled plot sprayer, *NC Weed Cont. Conf.*, 35.
Broadbent, L., Doncaster, J. P., Hull, R. and Watson, M. A. (1948) Equipment used for trapping and identifying aphids, *Proc. R. Ent. Soc. London (A)*, **23**, 57–8.
Brooks, T. W., Doane, C. C. and Haworth, J. K. (1979) Suppression of *Pectinophora gossypiella* with sex pheromones, *Proc. 1979 Br. Crop Prot. Conf. – Pests and Diseases*, **3**, 853–66.
Brown, E. S. (1965) Armyworm research, *E. Afr. Agric. For Res. Org. 1964*, 24–36.
Brown, E. S. and Mohammed, A. K. A. (1972) The relation between simulated armyworm damage and crop loss in maize and sorghum, *E. Afr. Ag. For J.*, **37**, 237–57.
Brown, E. S., Betts, E. and Rainey, R. C. (1969) Seasonal changes in distribution of the African armyworm *Spodoptera exempta* (Wlk) (*Lep. Noctuidae*), with special reference to eastern Africa, *Bull. Ent. Res.*, **58**, 661–728.
Brunskill, R. T. (1956) Factors affecting the retention of spray droplets on leaves, *Proc. 3rd Br. Weed Contr. Conf.*, **2**, 593–603.

Brunskill, R. T. (1957) A variable dosage sprayer for agricultural use, *J. Agric. Eng. Res.*, **2**, 135–40.

Burchfield, H. P., Hilchey, J. D. and Stores, E. E. (1952) An objective method for insecticide bioassay based on photomigration of mosquito larvae. *Contr. Boyce Thompson Inst.*, **17**, 57–86.

Burges, H. D. (1981) Safety, safety testing and quality control of microbial pesticides, in: H. D. Burges, (ed.) *Microbial Control of Pests and Plant Diseases 1970–1980*. Academic Press, London and New York, pp. 737–67.

Burges, H. D. and Hussey, N. W. (eds), (1971) *Microbial Control of Insects and Mites*. Academic Press, London and New York.

Burgess, M. W. (1983) Development of cotton-pest management in Zimbabwe, *Crop Prot.*, **2**, 247–50.

Burt, C. E., Smith, D. B. and Lloyd, E. P. (1966) A rotary disc device for applying ultra-low volume (undiluted) pesticides with ground equipment, *J. Econ. Ent.*, **59**, 1487–9.

Busvine, J. R. (1968) Detection and measurement of insecticide resistance in arthropods of agricultural or veterinary importance, *Wld. Rev. Pest Control*, **7**, 27–41.

Busvine, J. R. (1971) *A Critical Review of the Techniques for Testing Insecticides*. Commonwealth Inst. of Ent., London (345 pp.).

Busvine, J. R. (1980) *Recommended Methods for Measurement of Pest Resistance to Pesticides*. FAO Plant Production and Protection Paper No. 21 (132 pp.).

Byass, J. B., and Charlton, G. K. (1968) Equipment and methods for orchard spray application research. I: Laboratory application equipment, *J. Agric. Engng. Res.*, **13**, 280–9.

Byrde, R. J. W. (1981) Conducting and interpreting field fungicide trials, in: R. T. Burchill, (ed.), *Methods in Plant Pathology*. CAB Phytopathological Paper No. 26, pp. 38–43.

Caltagirone, L. E. (1981) Landmark examples in classical biological control, *Ann. Rev. Ent.* **26**, 213–32.

Campion, D. G. (1983) Pheromones for the control of insect pests in Mediterranean Countries. *Crop Prot.*, **2**, 3–16.

Campion, D. G., Bettany, B. W., Nesbitt, B. F., Beevor, P. S., Lester, R. and Poppi, R. G. (1974) Field studies of the female sex pheromone of the cotton leafworm *Spodoptera littoralis* (Boisd) in Cyprus, *Bull Ent. Res.*, **64**, 89–96.

Campion, D. G., Hunter-Jones, P., McVeigh, L. J., Hall, D. R., Lester, R. and Nesbitt, B. F. (1980) Modification of the attractiveness of the primary pheromone component of the Egyptian cotton leafworm *Spodoptera littoralis* (Boisd), (Lepidoptera, Noctuidae) by secondary pheromone components and related chemicals, *Bull. Ent. Res.*, **70**, 417–34.

Campion, D. G., McVeigh, L. J., Hunter-Jones, P., Hall, D. R., Lester, R., Nesbitt, B. F., Marrs, G. J. and Alder, M. R. (1980) Evaluation of micro-encapsulated formulations of pheromone components of the Egyption cotton leafworm *Spodoptera littoralis* (Boisd) in Crete, in: E. R. Mitchell (ed.), *Management of Insect Pests with Semiochemicals*. Plenum, New York.

Campion, D. G., McVeigh, L. J., Murlis, J., Hall, D. R., Lester, R., Nesbitt, B. F. and Marrs, G. J. (1976) Communication disruption of adult Egyptian cotton leafworm *Spodoptera littoralis* (Boisd) (Lepidoptera, Noctuidae) in Crete, using synthetic pheromones applied by micro-encapsulation and dispenser techniques, *Bull. Ent. Res.* **66**, 335–44.

Campion, D. G. and Nesbitt, B. F. (1981) Lepidopteran sex pheromones and pest management in developing countries, *Trop. Pest Man.*, **27**, 53–61.

Cannell, R. Q. (1981) Potentials and problems of simplified cultivation and conservation tillage, *Outlook on Agric.*, **10**, 379–84.

Cardarelli, N. F. (1976) *Controlled Release Pesticide Formulations*. CRC Press, Cleveland, Ohio, (210 pp.).
Cardarelli, N. F. (1979) In: H. Geissbühler (ed.) *Advances in Pesticide Science*. Vol. 3, pp. 744.
Cardarelli, N. F. and Cardarelli, B. M. (1982) Controlled release pesticides: a historical summary and state of the art, *Division of Isotope and Radiation. Application of Atomic Energy for Food and Agricultural Development*, FAO/IAEA Vienna.
Carino, F. O., Kenmore, P. E. and Dyck, V. A. (1979) The FARMCOP suction sampler for hoppers and predators in flooded rice fields, *Int. Rice Res. Newsl.*, **4**, 21–2.
Carson, A. G. (1978) Weeds and their control in Ghana, in: I. O. Akobundu, (ed.), *Weeds and their Control in the Humid and Subhumid Tropics*. IITA Proc. Ser. No. 3, pp. 150–8.
Carson, R. (1962) *Silent Spring*. Greenwick, CT: Fewcett.
Champ, B. R. and Dyte, C. E. (1976) *Report of the FAO Global Survey of Pesticide Susceptibility of Stored Grain Pests*. FAO Plant Production and Protection Ser.No.5.
Chant, D. A. (1962) A brushing method for collecting mites and small insects from leaves, *Prog. In Soil Zool.*, **1**, 222–5.
Chant, D. A. and Muir, R. C. (1955) A comparison of the imprint and brushing machine methods for estimating the number of the fruit tree red spider mite *Metatetranycus ulmi* (Koch) on apple leaves, Rep. E. Malling Res. Stn. 1954, 141–5.
Chapman, R. B. and Peerman, D. R. (1979) Negatively correlated cross-resistance to a synthetic pyrethroid in organophosphorus resistant *Tetranychus urticae, Nature*, **281**, 298–99.
Chiang, H. C. (1973) Bionomics of the northern and western corn rootworms, *Ann. Rev. Ent.*, **18**, 47–72.
Chiang, H. C. and French, L. K. (1980) Host tolerance, a short-term pest management tool – maize and corn rootworm as a model, *FAO Pl. Prot. Bull.*, **28**, 137–8.
Chiarappa, L. (ed.) (1971) *Crop Loss Assessment Methods*. FAO manual on the evaluation and prevention of losses by pests, diseases and weeds, CAB, Slough, UK.
Chiarappa, L. (ed.) (1981) *Crop Loss Assessment Methods*, Suppl. 3. CAB/FAO.
Chin, K. M. and Ajimilah, N. H. (1982) Rice variety mixtures in disease control, *Proc. Int. Conf. Pl. Prot. in Tropics, Kuala Lumpur*, 241–6.
Church, B. M. (1971) The place of sample survey in crop loss estimates, in: L. Chiarappa (ed.), *Manual on Crop Loss Assessment*. FAO Rome, 2.2/1–8.
Cialone, J. C. and Bayer, G. A. (1962) A logarithmic sprayer for small plots, *Proc. North Eastern Weed Cont. Conf.*, 77–81.
Cinch, P. G. (1971) A battery-operated vacuum device for collecting insects unharmed, *New Zealand Ent.*, **5**, 28–30.
Clare, R. W., Boothroyd, D. and Simkin, M. B. (1981) The cost effective use of fungicide in winter wheat in an area of high disease risk, *Proc. 1981 Br. Crop Prot. Conf. – Pests and Diseases*, 275–82.
Clarke, G. S. and Ross, A. A. (1964) A small-scale variable dosage (logarithmic sprayer), *Weed Res.*, **4**, 249–55.
Clarke, R. T. (1965) The presentation of numerical results of experiments for publication in a scientific agricultural journal, *Expl. Agric.*, **1**, 315–9.
Clausen, C. P. (ed.) (1978) *Introduced Parasites and Predators of Arthropod Pests and Weeds: a world review*. USDA ARS Handbook 480, Washington (545 pp.) (pp. 294–313).
Cleveland, M. L. (1962) Adhesives for holding mites to glass plates, *J. Econ. Ent.*, **55**, 570–1.

Coffee, R. A. (1979) Electrodynamic energy – a new approach to pesticide application, *Proc. 1979 Br. Crop Prot. Conf. – Pests and Diseases*, 777–89.
Coffee, R. A. (1980) Electrodynamic Spraying, *BCPC Monograph*, **24**, 95–107.
Coffee, R. A. (1981a) Recent developments in electrodynamic spraying, *Proc. 1981 Br. Crop Prot. Conf. – Pests and Diseases*, 883–9.
Coffee, R. A. (1981b) Electrodynamic crop spraying, *Outlook on Agriculture*, **10**, 350–6.
Coggins, S. and Baker, E. A. (1983) Microsprayers for the laboratory application of pesticides, *Ann. Appl. Biol.*, **102** 149–54.
Combellack, J. H. and Matthews, G. A. (1981) Droplet spectra measurements of fan and cone atomizers using a laser diffraction technique, *J. Aerosol Sci.*, **12**, 529–40.
Combellack, J. H. and Matthews, G. A. (1981) The influence of atomizer, pressure and formulation on the droplet spectra produced by high-volume sprayers, *Weed Res.*, **21**, 77–86.
Comins, H. N. (1977a) The development of insecticide resistance in the presence of migration, *J. Theor. Biol.*, **64**, 177–97.
Comins, H. N. (1977b) The management of pesticide resistance, *J. Theor. Biol.*, **65**, 399–420.
Comins, H. N. (1977c) *The Control of Adaptable Pests* in Proceedings of a Conference on Pest Management. (Ed. G. A. Norton and C. S. Holling), Laxenburg, Austria. 352 pp.
Conway. G. R. (1976) Man versus Pests in May, R. M. (ed.) *Theoretical Ecology: principles and applications* Blackwell Scientific Publications, Oxford, pp. 257–281.
Conway, G. R. (1977) Mathematical models in applied ecology, *Nature*, **269**, 291–7.
Conway, G. R., Norton, G. A., King, A. B. S. and Small, N. J. (1975) A systems approach to the control of the sugar cane froghopper, in: G. E. Dalton (ed.), *Study of Agricultural Systems*. Applied Science Publishers, London, pp. 193–229.
Cooke, B. K., Herrington, P. J. and Morgan, N. G. (1974) *Pesticide Fall-out and Orchard Earthworms*. Report, Long Ashton Research Station for 1974, 105–106.
Cramer, H. H. (1967) *Plant Protection and World Crop Production*. Bayer Pflanzenschutz Leverkusen (524 pp.).
Crane, E. and Walker, P. (1983) *The Impact of Pest Management on Bees and Pollination*. Report published by the International Bee Research Association and Tropical Development and Research Institute London, 129 pp. and 2 annexes.
Cranham, J. E. and Solomon, M. G. (1981) *Mite Management in Commercial Apple Orchards*. Report East Malling Research Station for 1980, pp. 171–2.
Croft, B. A. (1975a) Tree fruit pest management, in: R. L. Metcalf and R. W. H. Luckmann (eds), *Introduction to Insect Pest Management*. pp. 471–507.
Croft, B. A. (1975b) Integrated control of orchard pests in the USA, *Proc. 5th Sym. Int. Cont. in Orchards*. Bolzano OILB/SROP 1974, 109–24.
Croft, B. A. (1979) Management of apple arthropod pests and natural enemies relative to developed insecticide resistance, *Env. Ent.*, **8**, 583–6.
Croft, B. A. and Brown, A. W. A. (1975) Responses of arthropod natural enemies to insecticides, *Ann. Rev. Ent.*, **20**, 285–335.
Cronshaw, D. K. (19S2) Management of banana leaf spot (sigatoka) disease in the Windward Islands, *Trop. Pest Management*, **28**, 136–46.
Cueller, C. B. and Brinklow, D. M. (1973) The screw worm strikes back, *Nature*, **242**, 493–4.
Cunningham, J. C. and Entwistle, P. F. (1981) Control of sawflies by baculovirus, in: H. D. Burges, (ed.), *Microbial Control of Pests and Plant Diseases 1970–1980*. Academic Press, London and New York, pp. 379–407.

Curtis, C. F., Cook, L. M. and Wood, R. J. (1978) Selection for and against insecticide resistance and possible methods of inhibiting the evolution of resistance in mosquitoes, *Ecol. Ent.*, **3**, 273–86.

Cussans, G. and Moss, S. R. (1982) Population dynamics of annual grass weeds, in: R. G. Austin, (ed.), *Decision Making in the Practice of Crop Protection.* BCPC Monograph 25, 91–8.

David, W. A. L. (1945) Factors influencing the interaction of insecticidal mists and flying insects. I. The design of a spray testing chamber and some of its properties *Bull. Ent. Res.*, **36**, 373–393.

David, W. A. L. (1946) Factors influencing the interaction of insecticidal spray mists and their relation to flying insects, *Bull. Ent. Res*, **37**, 1–28.

Davidse, L. C., Looyen, D., Turkensteen, L. J. and Van Der Wal, D. (1981) Occurrence of metalaxyl – resistant strains of *Phytophthora infestans* in Dutch potato fields, *Netherlands Plant Pathology,* **87**, 65–68.

Davies, R. G. (1971) *Computer Programming in Quantitative Biology*. Academic Press, London and New York (Probit Analysis pp. 410–19).

Davis, H. (1967) *Tsetse Flies in Northern Nigeria.* Ibadan University Press, Ibadan.

Davis, J. M. (1951) A vibratory apparatus for producing droplets of a uniform size, *USDA Bureau of Entomology*, ET 295.

Day, B. E. (1958) A simplified logarithmic plot sprayer, *Weeds*, **6**, 441–6.

De Bach (1974) *Biological Control by Natural Enemies.* Cambridge University Press, Cambridge (323 pp.).

De Bach, P. and Bartlett, B. R. (1951) Effects of insecticides on biological control of insect pests of citrus, *J. Econ. Ent.*, **44**, 372–83.

De Bach, P. and Huffaker, C. B. (1971) Experimental techniques for evaluation of the effectiveness of natural enemies, in: C. B. Huffaker, (ed.), *Biological Control.* Plenum Press, New York, pp. 113–40.

De Bach, P., Rosen, D. and Kennett, C. E. (1971) Biological control of coccids by introduced natural enemies, in: C. B. Huffaker (ed.), *Biological Control.* Plenum, New York, pp. 165–94.

Delp, C. J. (1981) Strategies for dealing with fungicide resistance problems, *Proc. 1981 Br. Crop Prot. Conf. – Pests and Diseases*, 865–71.

Denholm, I. (1981) Present trends and future needs in modelling for the management of insecticide resistance, *Proc. 1981 Br. Crop Prot. Conf. – Pests and Diseases*, 847–55.

Denholm, I., Farnham, A. W., O'Dell, K. and Sawicki, R. M. (1983) Factors affecting resistance to insecticides in houseflies. *Musca domestica* L. (Diptera: Muscidae) I. Longterm control with bioresmethrin of flies with strong pyrethroid-resistance potential, *Bull. ent Res.*, **73**, 481–9.

Dewar, A. M., Woiwod, I. and Choppin de Janvry (1980) Aerial migrations of the rose-grain aphid *Metopolophium dirhodum* (Wlk) over Europe in 1979, *Pl. Path.*, **29**, 101–109.

Dietrick, E. J. (1961) An improved back-pack motor fan for suction trapping insect populations, *J. Econ. Ent.*, **54**, 394–5.

Dimmock, N. A. (1950) Production of uniform droplets. *Nature*, **166**, 686.

Dittrich, V. (1962) A comparative study of toxicological test methods on a population of the two-spotted spider mite (*Tetranychus telarius*), *J. Econ. Ent.*, **55**, 644–51.

Dittrich, V. (1981) Insecticide resistance – how can industry meet the challenge? *Proc. 1981 Br. Crop Prot. Conf. – Pests and Diseases*, 837–46.

Dombrowski, N. and Lloyd, T. L. (1974) Atomization of liquids by spinning cups, *Chem. Engng. J.*, **8**, 63–81.

Duffus, J. E. (1971) Role of weeds in the incidence of virus diseases, *Ann. Rev. Phytopath.*, **9**, 319–40.

Duncan, D. B. (1955) Multiple range and multiple F tests, *Biometrics*, **11**, 1–41.
Duncombe, W. C. (1973) The acaricide spray rotation for cotton, *Rhod. Agric. J.*, **70**, 115–18.
Duncombe, W. B. (1977) Cotton losses caused by spider mites (*Acarina tetranychidae*), *Rhod. Agric. J.*, **70**, 141–6.
Dyke, G. V. (1974) *Comparitive Experiments with Field Crops*. Butterworths, London.

Edwards, C. A. (1970) Effects of herbicide on the soil fauna, *Proc. 10th Br. Weed Cont. Conf.*, **3**, 1052–62.
Edwards, C. A. (1975) Effects of direct drilling on the soil fauna, *Outlook on Agriculture*, **8**, 243–4.
Edwards, C. A. and Steed, E. (1979) Annual Report, Rothamsted Experimental Station for 1978, 95–6.
Ehrenberg, A. S. C. (1978) Graphs or tables? *The Statistician*, **27**, 87–96.
Elliott, M. and Graham-Bryce, I. J. (1977) IRC, COPR, seminar report **3**, 4–7.
Elliott, M., Janes, N. F. and Potter, C. (1978) The future of pyrethroids in insect control, *Ann. Rev. Ent.*, **23**, 443–69.
Ennis, W. B. and Williamson, R. E. (1963) Influence of drop size on effectiveness of low-volume herbicidal spray, *Weeds*, **11**, 67–72.

Falcon, L. A. (1976) Problems associated with the use of arthropod viruses in pest control, *Ann. Rev. Ent.*, **20**, 97–117.
Farmery, H. (1978) Development of the Microdrop CDA Sprayer. *BCPC Monograph* 22, pp. 107–10.
Farrell, J. A. K. (1976a) Effects of groundnuts sowing date and plant spacing on rosette virus disease in Malawi, *Bull. Ent. Res.*, **66**, 159–72.
Farrell, J. A. K. (1976b) Effects of intersowing with beans on the spread of groundnut rosette virus by *Aphis craccivora* (Koch.) (Hemiptera, Aphididae) in Malawi, *Bull. Ent. Res.*, **66**, 331–4.
Finch, S. (1977) Monitoring insect pests of cruciferous crops, *Proc. 1977 Br. Crop Prot. Conf. – Pests and Diseases*, 219–26.
Findley, R. W. and Farber, D. A. (1981) *Environmental Law Cases and Materials*. West Publishing Co. St Paul, Minnesota.
Finney, D. J. (1949) The adjustment for a natural response rate in probit analysis, *Ann. Appl. Biol.*, **36**, 187–95.
Finney, D. J. (1955) *Experimental Design and its Statistical Basis*. Cambridge University Press, Cambridge.
Finney, D. J. (1971) *Probit Analysis* (3rd edn). Cambridge University Press. Cambridge.
Fisher, R. A. and Yates, F. (1964) *Statistical tables for biological, agricultural and Medical Research*. Longman, Harlow.
Flint, M. L. and van den Bosch, R. (1981) *Introduction to Integrated Pest Management*. Plenum, New York.
Foster, R. N., Staten, R. T. and Miller, E. (1977) Evaluation of traps for pink bollworm, *J. Econ. Ent.*, **70**, 289–91.
Franklin, M. T. (1970) Interrelations of nematodes, weeds, herbicides and crops, *Proc. 10th Br. Weed Cont. Conf.*, **3**, 927–33.
Franz, J. M. (1978) Report on the activities of the working group: pesticides and beneficial arthropods, *Bull. WPRS/IOBC, 4. 16*, 137–43.
Franz, J. M., Bogenschutz, H., Hassan, J. A., Huang, P., Nation, E., Sutter, H. and Viggias, G. (1980) Results of a joint pesticide test programme by the working group: pesticide and beneficial arthropods, *Entomophaga*, **25**, 231–6.
Franz, J. M., Hassan, S. A. and Bogenschutz, H. (1976) (Some results of standardized laboratory tests on the effect of pesticides on entomophagous beneficial arthropods), *Nach. Deutsch Pflanzenschutzd.*, **28**, 181–3.
Fraser, R. P. (1958) The fluid kinetics of application of pesticide chemicals, in:

R. L. Metcalfe (ed.), *Advances in pest control research*. Vol. II, Wiley Inter-Science, New York, pp. 1–106.

Freeman, G. H. (1967) Problems in plant pathology and entomology field trials, *Expl. Agric.*, **3**, 351–8.

Freeman, G. H. (1981) Design and analysis of field trials in plant pathology, in: R. T. Burchill, (ed.), *Methods in Plant Pathology*. CAB Phytopathological Paper No. 26, 32–7.

French P. (1980) A mechanized field sprayer for small plot pesticide trials, *Proc. 5th Int. Conf. on Mechanization of Field Experiments*, 135–42.

French, S. A. W. and Humphries, E. C. (1977) The effect of partial defoliation on yield of sugar beet, *Ann. Appl. Biol.*, **87**, 201–12.

Frost, A. R. (1981) Rotary atomization in the ligament formation mode, *J. Agric. Engng. Res.*, **26**, 63–78.

Frost, A. R. and Yates, W. E. (1981) A pulsed-jet atomizer for the production of sprays with narrow drop size ranges, *J. Agric. Engng. Res.*, **26**, 357–66.

Frost, S. W. (1952) *Light Traps for Insect Collection, Survey and Control*. Bulletin Pa. Agricultural Experimental Station 550 (32 pp.).

Fryer, J. D. (1956) A small-scale logarithmic sprayer, *Proc. 3rd Brit. Weed Control Conf.*, 585–90.

Fryer, J. D. (1977) Recent developments in the agricultural use of herbicides in relation to ecological effects, in: F. H. Perring, and K. Mellanby, (eds), *Ecological Effects of Pesticides*. Academic Press, London and New York, pp. 27–45.

Fryer, J. D. (1981) Weed management: fact or fable? *Phil. Trans. R. Soc.*, London, B 295, 185–97.

Fryer, J. D. and Elliott, J. G. (1954) Spraying equipment for the experimental application of herbicides, *Proc. 2nd Br. Weed. Cont. Conf.*, 375–88.

Fryer, J. D. and Makepeace, R. J. (eds) (1977) *Weed Control Handbook I: Principles*. Blackwell Scientific Publications, Oxford.

Funkhouser, W. A. and Las, A. S. (1981) Aerial dissemination of insect pheromones, *Ann. Soc. Ag. Eng. Paper* No. 81–1501.

Galley, D. J. (1968) A biological assay technique for the assessment and comparison of systemic insecticide residues, *Ann. Appl. Biol.*, **61**, 457–66.

Galley, D. J. and Gohain, R. (1981) The susceptibility of *Myzus persicae* and *Brevicoryne brassicae* to 14C-phorate in artificial diets, *Ann. Appl. Biol.*, **99**, 107–14.

Garnett, R. P. (1980) A low-volume herbicide applicator for tropical small-holder farmers, *Proc. 1980, Br. Crop Prot. Conf. – Weeds*, 629–36.

Gast, R. T. (1959) The relationship of weight of lepidopterous larvae to effectiveness of topically applied insecticides, *J. Econ. Ent.*, **52**, 1115–17.

Gebhardt, M. R. (1972) Gebhardt plot sprayer, in: ASAE/AMFE *Directory of North American Equipment for the Mechanization of Agricultural Research*. Aas-NLH, Norway, p. 82.

George, K. S. (1975) The establishment of economic damage thresholds with particular reference to cereal aphids, *Proc. 8th Br. Insectic. Fungic. Conf.* 79–85.

George, K. S. and Gair, R. (1979) Crop loss assessment on winter wheat attacked by the grain aphid *Sitobion avenae* (F.), 1974–1977, *Pl. Path.*, **28**, 143–9.

Georghiou, G. P. and Taylor, C. E. (1977) Genetic and biological influences in the evolution of insecticide resistance, *J. Econ. Ent.*, **70**, 319–23.

Gerber, H. R. (1977) Biotests for herbicide development, in: N. R. McFarlane, (ed.), *Crop Protection Agents: their biological evaluation*. Academic Press, London and New York, pp. 307–21.

Gillham, F. E. M. (1965) Evolutionary significance of glands and their importance in cultivated cotton, *Cott. Gr. Rev.*, **42**, 101–103.

Gledhill, J. A. (1982) Progress and problems in *Heliothis* management in tropical

Southern Africa, in: W. Reed (ed.), *Proceedings of International Workshop on Heliothis Management*. ICRISAT Patancheru, A.P. India, pp. 375–4.
Gomez, K. A. (1972) *Techniques for Field Experiments with Rice*. IRRI Philippines (46 pp.).
Gomez, K. A. and Gomez, A. A. (1976) *Statistical Procedures for Agricultural Research, with Emphasis on Rice*. IRRI Philippines (294 pp.).
Goodwin-Bailey, K. A., Holborn, J. M. and Davies, M. (1957) A technique for the biological evaluation of insecticidal aerosols, *Ann. Appl. Biol.*, **45**, 347–60.
Graham-Bryce, I. J. (1977) Crop protection: a consideration of the effectiveness and disadvantages of current methods and of the scope for improvement, *Phil. Trans. R. Soc. Lond.*, B 28, 163–79.
Graham-Bryce, I. J. (1981) The current status and future potential of chemical approaches to crop protection, *Phil. Trans. R. Soc. Lond.*, B, 295, 5–16.
Gratwick, M. (1963) A technique for exposing insects to insecticidal deposits on detached foliage, *Bull. Ent. Res.*, **54**, 81–2.
Gray, R. A., Green, L. L., Hock, P. E. and Pallos, F. M. (1982) The evaluation of practical crop safeners, *Proc. 1982 Br. Crop Prot. Conf. – Weeds*, 431–7.
Green, M. D., Hartley, G. S. and West, T. F. (1977) *Chemicals for Crop Protection and Pest Control*. Pergamon, Oxford.
Grosscurt, A. C. (1978) Diflubenzuron: some aspects of its ovicidal and larvicidal mode of action and an evaluation of its practical possibilities. *Pestic. Sci.*, **9**, 373–86.
Gudauskas, R. T. and Canerday, T. D. (1968) The effect of heat, buffer salt and H-ion concentration and ultraviolet light on the infectivity of *Heliothis* and *Trichoplusia* nuclear polyhedrosis viruses, *J. Invert. Path.*, **12**, 405–11.
Gutierrez, A. P., DeMichele, D. W., Wang, Y., Curry, G. L., Skeith, R. and Brown, L. G. (1980) The systems approach to research and decision making for cotton pest control, in: C. B. Huffaker (ed.), *New Technology of Pest Control*. Wiley Inter-Science, New York, pp. 155–86.

Hadaway, A. B. and Barlow, F. (1965) Studies on the deposition of oil drops, *Ann. Appl. Biol.*, **55**, 267–74.
Hagen, K. S., Sawall, E. F. and Tassan, R. L. (1971) The use of food sprays to increase effectiveness of entomophagous insects, *Proc. Tall Timbers Conf. Ecol. Animal Contr. Habitat Managem*, **2**, 59–81.
Hall, D. C., Norgard, R. B. and True, P. K. (1975) The performance of independent pest management consultants, *Calif. Agric.*, **29**, 12–14.
Hall, R. A. (1981) The fungus *Verticillium lecanii* as a microbial insecticide against aphids and scales, in: H. D. Burges (ed.), *Microbial Control of Pests and Plant Diseases 1970–1980*. Academic Press, London and New York, pp. 483–98.
Hammerton, J. L., Russell, R. L. and Springer, B. F. G. (1983) *Field Experimentation in the Tropics*. Longman, London.
Hanf, M. and Behrendt, S. (1977) Modern methods of assessing the performance and reliability of new herbicides, in: N. R. McFarlane (ed.), *Crop Protection Agents: their biological evaluation*. Academic Press, London and New York, pp. 323–32.
Hansberry, T. R. and Richardson, C. H. (1936) Toxicity of certain stomach poisons to several common lepidopterous larvae, *J. Econ. Ent.*, **29**, 1160–6.
Harding, W. C., Harsock, J. C. and Rohmer, G. G. (1966) Blacklight trap standards for general insect surveys, *Entom. Soc. Am. Bull.*, **12**, 31–2.
Hare, W. W. and Harrell, E. A. (1978) Field evaluation of experimental sprayer for control of insects on sweet corn, *J. Econ. Ent.*, **71**, 917–18.
Harlan, J. R. and Starks, K. J. (1980) Germplasm resources and needs, in: F. G. Maxwell and P. R. Jennings (eds), *Breeding Plants Resistant to Insects*. Wiley Inter-Science, New York, 253–73.
Harrap, K. A. (1982) Assessment of the human and ecological hazards of

microbial insecticides, *Parasitology*, **84**.

Harris, P. and Zwolfer, H. (1971) Technical Communication CIBC No. 4, 76–9.

Harrison, B. D. (1977) Ecology and control of viruses with soil inhabiting vectors, *Ann. Rev. Phytopath.*, **15**, 331–60.

Hartley, G. S. and Graham-Bryce, I. J. (1980) *Physical Principles of Pesticide Behaviour*. Academic Press, London and New York.

Hartstack, A. W. (1979) Light sources, trap design and the factors affecting moth catch, in: R. L. Rabb and G. G. Kennedy, (eds), *Movement of Highly Mobile Insects: concepts and methodology in research*, North Carolina State University, North Carolina, pp. 232–40.

Hartstack, A. W. (1982) Modelling and forecasting *Heliothis* populations in the USA, in: W. Reed (ed.), *Proceedings of the International Workshop on Heliothis Management*. ICRISAT, Patancheru, A.P. India, pp. 51–60.

Hartstack, A. W., Hollingsworth, J. P., Ridgway, R. L. and Coppedge, J. R. (1973) A population dynamics study of the bollworm and tobacco budworm with light traps, *Environ. Ent.*, **2**, 224–52.

Hartstack, A. W., Witz, J. A. and Buck, D. R. (1979) Moth traps for the tobacco budworm, *J. Econ. Ent.*, **72**, 519–22.

Hassan, S. A. (1977) Standardized technique for testing side-effects of pesticides on beneficial arthropods in the laboratory, *Z. Planzenkrankh. Pflanzensch.*, **84**, 158–63.

Hassall, K. A. (1972) *World Crop Protection* II: *Pesticides*. Iliffe, London.

Headley, J. C. (1972a) Defining the economic threshold, in: National Academy of Science, *Pest Control Strategies for the Future*, pp. 100–108.

Headley, J. C. (1972b) Economics of agricultural pest control, *Ann. Rev. Entomol.*, **17**, 273–86.

Heath, O. V. S. (170) *Investigation by Experiment*. Studies in Biology, No. 23, Arnold, London.

Heathcote, G. D. (1970) Weeds, herbicides and plant virus diseases, *Proc. 10th Br. Weed Cont. Conf.*, **3**, 934–41.

Heathcote, G. D. (1978) Review of losses caused by virus yellows in English sugar beet crops and the cost of partial control with insecticides, *Pl. Path.*, **27**, 12–17.

Heathcote, G. D., Palmer, J. M. P. and Taylor, L. R. (1969) Sampling for aphids by traps and crop inspection, *Ann. Appl. Biol.*, **63**, 155–66.

Heijne, C. G. (1978) A study of the effect of disc speed and flow rate on the performance of the 'Micron Battleship', *Proc. 1978, Br. Crop. Prot. Conf. – Weeds*, 673–9.

Heinrichs, E. A., Chelliah, S., Pickin, S. R., Valencia, S. L. Fabellar, L. T., Arceo, M. B. and Aquino, G. B. (1982) *A Manual of Methods on Insecticide Evaluation in Rice*. IRRI, Philippines.

Henderson, C. F. and McBurnie, H. V. (1943) *Sampling Technique for Determining Populations of Citrus Red Mite and its Predators*. USDA Circ. 671 (11 pp.).

Henneberry, T. J. and Howland, A. F. (1966) Response of male cabbage loopers to blacklight with or without the presence of the female sex pheromone, *J. Econ. Ent.*, **59**, 623–6.

Herrington, P. J., Mapother, H. R. and Stringer, A. (1981) Spray retention and distribution on apple trees, *Pestic. Sci.*, **12**, 515–20.

Hewlett, P. S. (1946) The design and performance of an atomizing nozzle for use with a spray tower for testing liquid insecticides. *Ann. Appl. Biol.*, **33**, 303–306.

Hewlett, P. S. (1947) A direct-spray technique for the biological evaluation of pyrethrum-in-oil insecticides for use against stored product insects in warehouses, *Ann. Appl. Biol.*, **34**, 357–74.

Hewlett, P. S. (1954) A micro-drop applicator and its use for the treatment of certain small insects with liquid insecticides, *Ann. Appl. Biol.*, **41**, 45–64.

Hewlett, P. S. (1962) Toxicological studies on a beetle, *Alphitobius laevigatus* (F.) 1. Dose-response relations for topically applied solution of four toxicants in a

non-volatile oil, *Ann. Appl. Biol.*, **50**, 335–49.
Hewlett, P. S. and Lloyd, C. T. (1959) Topical application of mobile liquids to insects by means of micro-capillary tubes, *Ann. Appl. Biol.*, **48**, 125–33.
Hewlett, P. S. and Plackett, R. L. (1979) *The Interpretation of Quantal Responses in Biology*. Arnold, London.
Hewson, R. T. and Roberts, H. A. (1973) Some effects of weed competition on the growth of onions, *J. Hort. Sci.*, **48**, 51–7.
Himel, C. M. (1969a) The fluorescent particle spray droplet tracer method, *J. Econ. Ent.*, **62**, 912–6.
Himel, C. M. (1969b) The optimum size for insecticide spray droplets, *J. Econ. Ent.*, **62**, 919–25.
Hind, N. J. (1978) The Richmond Gibson Controlled Drop Application Trials Sprayer. *BCPC Monograph* 22, pp. 117–19.
Hinze, J. O. and Milborn, H. (1950) Atomization of liquids by means of a rotating cup, *J. Appl. Mech.*, **17**, 145–55.
Hirano, S. and Kira, T. (1965) Influence of autotoxic root exudation on the growth of higher plants grown at different densities (Intraspecific competition among higher plants XII), *J. Biol.*, Osaka City, University 16, 27–44.
Hirao, J. (1979) Forecasting brown planthopper outbreaks in Japan, in: *Brown Planthopper – Threat to Rice Production in Asia*. IRRI, pp. 101–12.
Hirst, J. M. R. (1952) An automatic volumetric spore trap, *Ann. Appl. Biol.*, **39**, 257–65.
Hislop, R. G. and Prokopy, R. J. (1981) Integrated management of phytophagous mites in Massachusetts (USA) apple orchards II: Influence of pesticides on the predator *Amblyseius fallacis* (Acarina Phytoseiidae) under laboratory and field conditions, *Prot. Ecol.*, **3**, 157–72.
Ho, S. H. and Galley, D. J. (1982) Relative toxicities of phorate and some of its metabolites to *Aphis fabae* (Scop.), *Pestic. Sci.*, **13**, 183–88.
Hollingsworth, J. P., Harstock, J. G. and Stanley, J. M. (1963) *Electric Insect Traps for Survey Purposes*. USDA ARS 42–3–1 (24 pp.).
Holly, K. (1982) The evaluation of a new herbicide, in: H. A. Roberts, (ed.) *Weed Control Handbook*: *Principles*, pp. 219–67.
Hughes, R. D. (1960) A method of estimating the numbers of cabbage root fly pupae in the soil, *Plant Path.*, **9**, 15–17.
Hussey, N. W. and Bravenboer, L. (1971) Control of pests in glass-house culture by the introduction of natural enemies, in: C. B. Huffaker (ed.), *Biological Control*. pp. 195–216. Plenum Press, New York.
Hussey, N. W. and Parr, W. J. (1963) The effect of glasshouse red spider mite (*Tetranychus urticae* Koch) on the yield of cucumbers, *J. Hort. Sci.*, **38**, 255–63.

Ingram, W. R. and Green, S. M. (1972) Sequential sampling for bollworms on raingrown cotton in Botswana, *Cott. Gr. Rev.*, **49**, 265–75.
Ivens, G. W. (1979) Nodding thistles, *New Zealand J. Agric.*, **138**, 16–17.

Jackson, T. A. (1980) The effect of defoliation on yield of radish, *Ann. Appl. Biol.*, **94**, 415–19.
Jaques, R. P. (1972) Control of the cabbage looper and the imported cabbage worm by viruses and bacteria, *J. Econ. Ent.*, **65**, 757–60.
Jellis, G. J. (1974) Common scab (*Steptomyces scabies*), Annual Report, Plant Breeding Institute (1973), Cambridge, 118.
Jenkyn, J. F. (1974) Comparison of seasonal changes in deposition of spores of *Erysiphe graminis* on different trapping surfaces, *Ann. Appl. Biol.*, **76**, 257–67.
Jenkyn, J. F. (1981) Experimental design and interplot interferance , in: L. Chiarappa (ed.), *Crop Loss Assessment Methods Suppl.*, 3, pp. 35–41.
Johnson, C. G. (1950) A suction trap for small airborne insects which automatically segregates the catch into successive hourly samples, *Ann. Appl. Biol.*, **37**, 80–91.

Johnson, C. G. and Taylor, L. R. (1955) The development of large suction traps for airborne insects, *Ann. Appl. Biol.*, **43**, 51–61.

Johnson, C. G., Southwood, T. R. E. and Entwistle, H. M. (1957) A new method of extracting arthropods and molluscs from grassland and herbage with a suction apparatus, *Bull. Ent. Res.*, **48**, 211–18.

Johnson, R. and Law, C. N. (1975) Genetic control of durable resistance to yellow rust (*Puccina striiformis*) in the wheat cultivar Hybride de Bersee, *Ann. Appl. Biol.*, **81**, 385–91.

Johnstone, D. R. (1970) A laboratory method for applying low volume spray patterns of variable form and density, *J. Agric. Eng. Res.*, **15**, 188–93.

Johnstone, D. R. (1973) Insecticide concentration for ultra-low-volume crop spray applications *Pestic. Sci.*, **4**, 77–82.

Judenko, E. (1969) A method of assessing losses in spring oats from fritfly (*Oscinella frit* (L.)) shoot attack, *Bull. Ent. Res.*, **59**, 479–84.

Judenko, E. (1973) Analytical method for assessing yield losses caused by pests on cereal crops with and without pesticides, *Trop. Pest Bull. No. 2*.

Joyce, R. J. V. and Beaumont, J. (1978) Collection of spray droplets and chemical by larvae, foliage and ground deposition, in: A. V. Holden, and D. Bevan, (eds), *Control of Pine Beauty Moth by Fenitrothion in Scotland 1978*. Forestry Commission, UK (176 pp.).

Joyce, R. J. V. and Roberts, P. (1959) The determination of the size of plot suitable for cotton spraying experiments in the Sudan El Gezira, *Ann. Appl. Biol.*, **47**, 287–305.

Kanno, H., Ikeda, K., Asai, T. and Mackawa, S. (1981) 2-tert-butylimino-3-isopropyl-5-phenylperhydro-1, 3, 5-thiadiazin-4- one (NN1-750), a new insecticide, *Proc. 1981 Brit. Crop Prot. Conf. – Pest and Diseases*, 59–66.

Kajihara, O., Asai, T., Ikada, K. and Lim, S. S. (1982) Buprofezin a new insecticide for control of brown planthopper, *Nilapavata lugens, Proc. Int. Conf. Pl. Prot. in Tropics*, Kuala Lumpur, 711 (abstract only).

Kasasian, L. and Seeyave, J. (1969) Critical periods for weed competition, *PANS* **15**, 208–12.

Kehat, M., Gothilf, S., Dunkelblum, E. and Greenberg, S. (1980) Field evaluation of female sex pheromone components of the cotton bollworm. *Heliothis armigera* (Hubner), *Ent. Exp. and Appl.*, **27**, 188–93.

Kennedy, J. S., Ainsworth, M. and Toms, B. A. (1948) Laboratory studies on the spraying of locusts at rest and flight, *Anti-locust Bull.*, **2** (64 pp.).

Kettle, D. S. (1949) The speed of action of insecticidal sprays and deposits and its use in assessing the biological efficiency of BHC, DDT and pyrethrum, *Bull. Ent. Res.*, **40**, 403–28.

Khosla, R. K. (1977) Techniques for assessment of losses due to pests and diseases of rice, *Ind. J. Ag. Sci.*, **47**, 171–4.

Kiritani, K. (1977) A systems approach to pest management of the green rice leafhopper, in: G. A. Norton and C. S. Holling (eds), *Proc. Conf. on Pest Management*. IIASA, Laxenburg, Austria, pp. 229–52.

Kisimoto, R. (1979) Brown planthopper migration, in: *Brown Planthopper Threat to Rice Production in Asia*. IRRI, pp. 113–24.

Knipling, E. F. (1979) *The Basic Principles of Insect Population Suppression and Management*. USDA Agriculture Handbook No. 512 (659 pp.).

Kogan, M. (1976) Evaluation of economic injury levels for soyabean insect pests, in: L. D. Hill (ed.), *World Soyabean Research*. Proc. Soyabean Research Conference September 1975, Champaign, Illinois, pp. 515–33.

Kogan, M., Ruesink, W. G. and McDowell, K. (1974) Spatial and temporal distribution patterns of the bean leaf beetle *Cerotoma trifurcata* (Forster), on soybeans in Illinois, *Environ. Ent.*, **3**, 607–17.

Koopman, M. J. and Balk, W. (1973) A simple laboratory spray method for use in crop protection research with special reference to ULV practice, *Pestic. Sci.*, **4**, 435–45.

Lake, J. R. and Taylor, W. A. (1974) Effect of the form of a deposit on the activity of barban applied to *Avena fatua (L.)*, *Weed Res.*, **14**, 13–18.

Landis, B. J. (1972) The alighting response of aphids to yellow pan water traps at different elevations, *Env. Ent.*, **1**, 473–6.

Large, E. C. (1958) Losses caused by potato blight in England and Wales, *Pl. Path.*, **7**, 39–48.

Law, S. E. (1980) Droplet charging and electrostratic deposition of pesticide sprays – research and development in its USA. *BCPC Monograph*, **24**, 85–94.

Lawson, T. J. and Uk, S. (1978) The Effect of Wind Turbulence and Crop Characteristics on the Dispersal of Aerial Sprays. *BCPC Monograph* 22, pp. 67–74.

Lawson, T. J. and Uk, S. (1979) The influence of wind turbulence, crop characteristics and flying height on the dispersal of aerial sprays, *Atmos. Ev.*, **13**, 711–15.

Leasure, J. K. (1959) A logarithmic–concentration sprayer for small plot use, *Weeds*, **7**, 91–97.

Le Clerg, E. (1971) Field experiments for assessment of crop losses, in: L. Chiarappa (ed.), *Manual on Crop Loss Assessment*. FAO, Rome, 211/1–11.

Ledieu, M. S. (1979) Laboratory and glasshouse screening of pesticides for adverse effects on the parasite *Encarsia formosa* Gahan, *Pestic. Sci.*, **10**, 123–32.

Legge, J. B. B. (1966) Behaviour of *Myzus persicae* (Sulz) on tobacco. The effects of time of planting and the pattern of growth of the plants on infestation in the field, *Rhod. Zambia Malawi. J. Agric. Res.*, **4**, 29–33.

Lewis, F. H. (1978) Field plot Tests of Apple Fungicides in *Methods for Evaluating Plant Fungicides, Nematicides and Bactericides*, The American Phytopathological Society, 40–43.

Lewis, T. (1969) The distribution of flying insects near a low hedgerow, *J. Appl. Ecol.*, **6**, 443–52.

Lewis, T. (1977) Prospects for monitoring insects using behaviour-controlling chemicals, *Proc. 1977, Br. Crop. Prot. Conf. – Pests and Diseases*, **3**, 847–56.

Lewis, T. (1981) Pest monitoring to aid insecticide use, *Phil. Trans. R. Soc.*, Lond. B295, 153–62.

Lewis, T. and Macauley, E. D. M. (1976) Design and evaluation of sex attractant traps for pea moth *Cydia nigricana* (Steph) and the effect of plume shape on catches, *Ecol. Ent.*, **1**, 175–87.

Lewis, T. and Sturgeon, D. M. (1978) Early warning of egg hatching in pea moth (*Cydia nigricana*), *Ann. Appl. Biol.*, **88**, 199–210.

Lewis, W. J., Hagen, K. S., Roelofs, W. L. and Schoonhoven, L. M. (1980) Status and potential use of behavioural chemicals in pest management, *FAO Pl. Prot. Bull.*, **28**, 121–8.

Lincoln, C., Boyer, W. P., Dowell, G. C., Barnes, G. and Dean, G. (1970) *Six Years Experience with Point – Sample Cotton Insect Scouting*. Agricultural Experimental Station, University of Arkansas Bull. 754 (30 pp.).

Lofgren, C. S., Ford, H. R., Tonn, R. J. and Jatansen, S. (1970) The effectiveness of ultra-low volume applications of Malathion at the rate of 6 US fluid ounces per acre in controlling *Aedes aegypti* in a large-scale test at Nakhon Sawan, Thailand, *Bull. Wld. Hlth. Org.*, **42**, 15–25.

Luck, R. F., Allen, J. C. and Baasch, D. (1980) A systems approach to research and decision-making in the citrus ecosystem, in: C. B. Huffaker, (ed.), *New Technology of Pest Control*. Wiley Inter-Science, New York, pp. 365–96.

Lukefahr, M. J. and Houghtaling, J. E. (1969) Resistance of cotton strains with high gossypol content to *Heliothis* spp., *J. Econ. Ent.*, **62**, 588–91.

Lutman, P. J. W. (1980) A Review of Techniques that Utilize Height Differences Between Crops and Weeds to Achieve Selectivity. *BCPC Monograph* 24, pp. 291–7.

Lutman, R. P. W. (1979) The effect of topical application of glyphosate and amino triazole on volunteer potatoes (*Solanum tuberosum*), *Weed Res.*, **19**, 377–84.

Macaulay, E. D. M. (1977) Field trials of an early-warning system for pea moth sex attractant, *Plant Path.*, **26**, 179–88.

MacCuaig, R. D. (1962) The collection of spray droplets by flying locusts, *Bull. Ent. Res.*, **53**, 111–23.

MacCuaig, R. D. and Watts, W. S. (1968) A simple technique for applying small measured quantities of insecticides to insects, *Bull. Ent. Res.*, **57**, 549–52.

McKinley, K. S., Brandt, S. A., Morse, P. and Ashford, R. (1972) Droplet size and phytotoxicity of herbicides, *Weed Sci.*, **20**, 450–2.

McKinley, K. S. and Glen, G, S. (1971) Shrouded spinning disc for the production of homogeneous sprays, *Can. Agric. Eng.*, **13**, 19–22.

McVeigh, L. J., Paton, E. M. and Hall, D. R. (1979) Factors affecting the performance of pheromone traps for male *Spodoptera littoralis. Proc. 1979 Br. Crop Prot. Conf. – Pests and Diseases*, 409–19.

McWilliams, J. M. (1979) *Heliothis* spp. Eggs: spray application to establish larvae infestations in experimental field plots of soyabeans, *Environ. Ent.*, **8**, 1044–6.

Marchant, J. A. and Green, R. (1982) An electrostatic charging system for hydraulic spray nozzles, *J. Agric. Eng. Res.*, **27**, 309–19.

Marks, R. J. (1976a) Field studies with the synthetic sex pheromone and inhibitor of the red bollworm *Diparopsis castanea* (Hmps) (Lepidoptera, Noctuidae) in Malawi, *Bull. Ent. Res.*, **66**, 243–65.

Marks, R. J. (1976b) Field evaluation of gossyplure; the synthetic sex pheromone of *Pectinophora gossypiella* (Saund) (Lepidoptera, Gelechiidae) in Malawi, *Bull. Ent. Res.*, **66**, 267–78.

Marks, R. J. (1977) Assessment of the use of sex pheromone traps to time chemical control of red bollworm *Diparopsis castanea* (Hmps) (Lepidoptera, Noctuidae) in Malawi, *Bull. Ent. Res.*, **67**, 575–87.

Marks, R. J. (1978) The influence of pheromone trap design and placement on catch of the red bollworm of cotton *Diparopsis castanea* (Hmps) (Lepidoptera, Noctuidae), *Bull. Ent. Res.*, **68**, 31–45.

Marks, R. J., Nesbitt, B. F., Hall, D. R. and Lester, R. (1978) Mating disruption of the red bollworm of cotton *Diparopsis castanea* (Hmps) (Lepidoptera, Noctuidae) by ultra-low volume spraying with a micro-encapsulated inhibitor of mating, *Bull. Ent. Res.*, **68**, 11–29.

Martin, H. (1973) *The Scientific Principles of Crop Protection*. Arnold, London (423 pp.).

Mason, B. J., Jayaratne, O. W. and Woods, J. D. (1963) An improved vibrating capillary device for producing uniform water droplets of 15–500 μm radius, *J. Sci. Instrum.*, **40**, 247–9.

Matthews, G. A. (1966) Investigations of the chemical control of insect pests of cotton in Central Africa. I: Laboratory rearing methods and tests of insecticides by application to bollworm eggs, II: Tests of insecticides with larvae and adults, *Bull. Ent. Res.*, **57**, 69–91.

Matthews, G. A. (1972) Response of cotton to a fertilizer mixture in Malawi, *Cott. Gr. Rev.*, **49**, 28–38.

Matthews, G. A. (1973a) Ultra-low volume spray application on cotton in Malawi, PANS **19**, 48–53.

Matthews, G. A. (1973b) Ultra-low volume spraying of cotton in Malawi, *Cott. Gr. Rev.*, **50**, 242–67.

Matthews, G. A. (1979a) *Pesticide Application Methods*. Longman, London.
Matthews, G. A. (1979b) Equipment for controlled droplet application of pesticides, *Int. Pest. Cont.*, **21**, 139–42.
Matthews, G. A. (1981) Improved systems of pesticide application, *Phil. Trans. R. Soc.*, Lond., B 295, 163–73.
Matthews, G. A., Rowell, J. G. and Beeden, P. (1972) Yield and plant development of reduced cotton stands in Malawi, *Exp. Agric.*, **8**, 33–48.
Matthews, G. A. and Tunstall, J. P. (1967) Insect attack and crop loss on cotton in Rhodesia, *Cott. Gr. Rev.*, **44**, 269–83.
Matthews, G. A. and Tunstall, J. P. (1968) Scouting for pests and the timing of spray application, *Cott. Gr. Rev.*, **45**, 115–27.
May, K. R. and Clifford, R. (1967) The impaction of aerosol particles on cylinders, spheres, ribbons and discs. *Am. Occp. Hyg.*, **10**, 83–95.
Mercer, W. B. and Hall, A. D. (1911) The experimental error of field trials, *J. Agric. Sci.*, **4**, 107.
Merritt, C. R. (1980) The Influence of Application Variables on the Biological Performance of Foliage Applied Herbicides. *BCPC Monograph* 24, pp. 35–43.
Merritt, C. R. (1982a) The influence of form of deposit on the phytotoxicity of difenzoquat applied as individual drops to *Avena fatua* (L.), *Ann. Appl. Biol.*, **101**, 517–25.
Merritt, C. R. (1982b) The influence of form of deposit on the phytotoxicity of MCPA, Paraquat and Glyphosate applied as individual drops, *Ann. Appl. Biol.*, **101**, 527–32.
Merritt, C. R. and Drinkwater, J. A. (1977) A laboratory device for producing single drops in uniform sizes of 150–450 m diameter, *Pestic. Sci.*, **8**, 457–62.
Metcalf, R. L. (1975) Insecticides in pest management, in: R. L. Metcalf and W. Luckman, (eds), *Introduction to Insect Pest Management*. Wiley Inter-Science, New York, pp. 235–73.
Metcalf, R. L. (1980) Changing role of insecticides in crop protection, *Ann. Rev. Ent.*, **25**, 219–56.
Metcalf, R. L. and Luckmann, W. (1975) *Introduction to Insect Pest Management*, Wiley Inter-Science, New York.
Miles, P. W. (1968) Insect secretions in plants, *Ann. Rev. Phytopath.*, **6**, 137–64.
Mitchell, R. S. and Rose, D. J. W. (1979) Factors affecting fluctuations in extent of *Salvinia molesta* on Lake Kariba, *PANS*, **25**, 171–77.
Mittler, T. E. and Dadd, R. H. (1964) An improved method for feeding aphids *Myzus persicae* (Sulz) on artificial diets, *Ann. Ent. Soc. Amer.*, **57**, 139–40.
Miyata, T., Sakai, H., Saito, T., Yoshioka, K., Tsuboi, A., Ozaki, K. and Saki, Y. (1979) Synergistic action of Malathion and Kitazin P in green rice leafhoppers and mice, in: *Annual Meeting of the Japanese Society of Applied Entomology and Zoology* (cited in Ozaki 1981, op. cit.).
Moore, N. W. (1977) The future prospect for wildlife, in: F. H. Perring, and K. Mellanby, (eds), *Ecological Effects of Pesticides*. Academic Press, London and New York, pp. 175–80.
Mortimer, A. M., McMahon, D. J., Manlove, R. J. and Putwain, P. D. (1980) The prediction of weed infestations and cost of differing control strategies, *Proc. Br. Crop Prot. Conf. – Weeds*, 415–22.
Morton, N. (1982) The 'Electrodyn' sprayer: first studies of spray coverage on cotton, *Crop Prot.* **1**, 27–54.
Moss, S. R. (1981) Techniques for the assessment of *Alopecurus myosuroides, Grass Weeds in Cereals in the UK Conf. 1981*, 101–107.
Mowlam, M. D., Nyirenda, G. K. C. and Tunstall, J. P. (1975) Ultra-low volume application of water-based formulations of insecticide to cotton, *Cott. Gr. Rev.*, **52**, 360–70.
Mukerjea, T. D. (1953) The relationship between the stage of development and

susceptibility to DDT and the pyrethrins of *Diataraxia oleracea* (L.), *Tenebrio molitor* (L.) and *Periplaneta americana* (L.), *Bull. Ent. Res.*, **44**, 121–61.

Mulder, R. and Gijswijt, M. J. (1973) The laboratory evaluation of two promising new insecticides which interfere with cuticle deposition, *Pestic. Sci.*, **4**, 737–45.

Mulhern, T. D. (1942) *New Jersey Mechanical Trap for Mosquito Surveys*. New Jersey Agricultural Experimental Station Circular 421, pp. 1–8.

Mumford, J. D. (1977) Farmer attitudes towards the control of aphids on sugar beet, *Proc. 1977 Br. Crop Prot. Conf. – Pests and Diseases*, **263–70.**

Mumford, J. D. (1981) Pest control decision making: sugar beet in England, *J. Ag. Econ.*, **32**, 31–41.

Munthali, D. C. (1981) *Biological Efficiency of Small Pesticide Droplets*. Ph.D. thesis, University of London.

Munthali, D. C. and Scopes, N. E. A. (1982) A technique for studying the biological efficiency of small droplets of pesticide solutions and a consideration of its implications, *Pestic. Sci.*, **13** (1), 60–3.

Needham, P. H. and Devonshire, A. L. (1973) A technique for applying small drops of insecticide solution to *Myzus persicae* (Sulz), *Pestic. Sci.*, **4**, 107–111.

Nemec, S. J. (1971) Effects of lunar phases on light-trap collections and populations of bollworm moths, *J. Econ. Ent.*, **64**, 860–4.

Nesbitt, B. F., Beevor, P. S., Hall, D. R. and Lester, R. (1980) (Z)-9-hexadecenal: a minor component of the female sex pheromone of *Heliothis armigera* (Hubner) (Lepidoptera, Noctuidae), *Ent. Exp. and Appl.*, **27**, 306–308.

Nightingale, W. H. (1979) Improved means of obtaining sustained uniform emission of bioactive substances, *Proc. 1979, Br. Crop Prot. Conf. – Pests and Diseases*, 401–407.

Norton, G. A. (1977) Background to agricultural pest management modelling, in: G. A. Norton and C. S. Holling (eds), *Proc. Conf. on Pest Management.* IIASA, Laxenburg, Austria, pp. 161–76.

Norton, G. A. (1979) Systems analysis and pest management – a pragmatic approach, *Proc. of the 3rd Australian Applied Entomology Conference*, Gatton 1977, 17–37.

Norton, G. A., Sutherst, R. W. and Maywald, G. F. (1983) A framework for integrating control methods against the cattle tick *Boophilus microplus* in Australia, *J. Appl. Ecol.*, **20**, 489–505.

Ooi, P. A. C. (1982) A surveillance system for rice planthoppers in Malaysia, *Proc. Int. Conf. Pl. Prot. in Tropics*, Kuala Lumpur, 551–65.

Ozaki, K. (1980) Resistance of rice insect pests to insecticides in Japan. Presented at the XVI International Congress of Entomology, Kyoto, 3–9 August.

Ozaki, K. (1981) Suppression of resistance through synergistic combination with emphasis on planthoppers and leafhoppers infesting rice in Japan, in: G. P. Georghiou and T. Saito (eds), *Pest Resistance to Pesticides: challenges and prospects*. (in press).

Painter, R. H. (1951) *Insect Resistance in Crop Plants*. MacMillan, New York.

Parham, M. R. (1982) Weed control in arable crops with the 'Electrodyn' sprayer, *Proc. 1982 Br. Crop Prot. Conf. – Weeds*, 1017–23.

Parker, C. (1978) Parasitic weeds and their control in the tropics, in: I. O. Akobundu, (ed.), *Weeds and their Control in the Humid and Subhumid Tropics*. IITA Proc. No. 3, pp. 22–50.

Parker, C., Richardson, W. G. and West, T. M. (1980) The potential for extending the selectivity of DPX 4189 by the use of herbicide safeners, *Proc. Br. Crop. Prot. Conf. – Weeds*, 15–22.

Parker, R. E. (1973) *Introductory Statistics for Biology*. Arnold, London.

Parkin, C. S. and Wyatt, J. C. (1980) The Measurement of Drop Spectra in Agricultural Sprays using a Particle Measuring Systems Optical Array Spectrometer. *BCPC Monograph* 24, pp. 241–9.

Parnell, F. R. (1925) Breeding jassid resistant cottons, *Emp. Cott. Gr. Rev.*, **2**, 330–6.

Parnell, F. R., King, H. E. and Ruston, D. F. (1949) Jassid resistance and hairiness of the cotton plant, *Bull. Ent. Res.*, **39**, 539–75.

Parr, H. C. M. and Busvine, J. R. (1948) A spinning-disc sprayer for applying residual insecticides, *Ann. Appl. Biol.*, **35**, 359–68.

Parr, W. J. (1968) Biological Control of Greenhouse whitefly Trialeurodes vaporariorum by the Parasite *Encarsia formosa* on Tomatoes. Report, Glasshouse Crops Research Institute (1966) pp. 135–9.

Pathak, M. D. (1970) Genetics of plants in pest management, in: R. L. Rabb and F. E. Guthrie (eds), *Concepts of Pest Management*, North Carolina State University, North Carolina, pp. 138–57.

Payne, C. C. (1982) Insect viruses as control agents, *Parasitology*, **84**, 35–77.

Pearce, S. C. (1976) *Field Experimentation with Fruit Trees and Other Perennial Plants*. CAB Technical Communication No. 23, (rev.).

Phillips, F. T., Etheridge, P. and Scott, G. C. (1976) Formulation and field evaluation of experimental baits for the control of leaf-cutting ants (Hymenoptera, Fornicidae), *Bull. Ent. Res.*, **66**, 579–85.

Phillips, J. R. (1978) Integrated insect management systems, in: *Sym. Dev. Opt. Crop Prod. Systems for the Mid South*, University of Arkansas, special report 67, pp. 104–105.

Phillips, J. R., Gutierrez, A. P. and Adkisson, P. L. (1980) General accomplishments towards better insect control in cotton, in: C. B. Huffaker, (ed.), *New Technology of Pest Control*. Wiley Inter-Science, New York, pp. 123–53.

Pimentel, D. (1961) On a genetic feed-back mechanism regulating populations of herbivores, parasites and predators, *Am. Nat.*, **95**, 65–79.

Plattner, H. C. (1975) (Studies on the biology of *Phygadeuon trichops* (Thoms.) (Hym.) (Ichneumonidae), *Anz. Schadlingskde pflanzensch*, Umweltsch., **48**, 56–60.

Plowman, R. E., Stonebrioge, W. C. and Hawtree, J. N. (1980) Fluazifop butyl – A new selective herbicide for the control of annual and perennial grass weeds. *Proc. Br. weed Cont. Conf.*, **1**, 29–38.

Polley, R. W. and Clarkson, J. D. S. (1978) Forecasting cereal disease epidemics, in: P. R. Scott and A. Bainbridge (eds), *Plant Disease Epidemiology*. Blackwell, Oxford, pp. 141–50.

Potter, C. (1941) A laboratory spraying apparatus and technique for investigating the action of contact insecticides with some notes on suitable test insects, *Ann. Appl. Biol.*, **28**, 142–69.

Potter, C. (1952) An improved laboratory apparatus for applying direct sprays and surface films, with data on the electrostatic charge on atomized spray fluids, *Ann. Appl. Biol.*, **39**, 1–28.

Potts, G. R. and Vickerman, G. P. (1974) Studies on the cereal ecosystem, *Adv. Ecol. Res.*, **8**, 107–87.

Pradhan, S. (1949) Studies on the toxicity of insecticide films 1: Preliminary investigations on concentration–time–mortality relation, *Bull. Ent. Res.*, **41**, 1–25.

Quick, G. R. and Buckele, W. F. (1972) *ASAE/IAMFE Directory of North American Equipment for the Mechanization of Agricultural Research*. ASAE and IAMFE, Aas NLH, Norway (210 pp.).

Reed, W. (1972) Uses and abuses of unsprayed controls in spraying trials, *Cott. Gr. Rev.*, **49**, 67–72.

Reed, W. (1981) Pigeon pea – Entomology, in: ICRISAT Annual Report for 1979–80. Patancheru, A.P. India, p. 113.

Rice, E. L. (1974) *Allelopathy*. Academic Press, London and New York.

Richardson, M. J. (1981) The determination of yield-loss relationships by non-plot methods, in: L. Chiarappa (ed.), *Crop Loss Assessment Methods*, Suppl. 3. CAB/FAO. pp. 59–63.

Richardson, R. H. (ed.) (1978) *The Screwworm Problem: evolution of resistance to biological control*. Texas Press.

Ridgman, W. J. (1969) Presentation of results from factorial experiments, *Cott. Gr. Rev.*, **46**, 288–300.

Rietz, S. and Kohsiek, H. (1982) A new developed plot sprayer. Gesunde Pflanzen, **34**, 84–8.

Rijsdijk, F. H. (1982) The EPIPRE system, in: R. B. Austin (ed.) *Decision making in the Practice of Crop Protection*. BCPC Monograph 25, pp. 65–76.

Rijsdijk, F. H. and Zadoks, J. C. (1979) A data bank on crop losses: first experience, *EPPO Bull.*, **9**, 297–303.

Ripper, W. E. (1944) Biological control as a supplement to chemical control of insect pests, *Nature*, **153**, 448–52.

Roberts, H. A. (1976) Weed competition in vegetable crops, *Ann. Appl. Biol.*, **83**, 321–47.

Robinson, H. S. and Robinson, R. J. M. (1950) Some notes on the observed behavior of Lepidoptera in flight in the vicinity of light sources together with a description of a light trap designed to take entomological samples, *Entomol. Gaz.*, **1**, 3–20.

Rohde, G. R. and Fay, P. K. (1980) A versatile tractor sprayer for herbicide research, *Proc. Western Soc. Weed Sic.*, **33**.

Room, P. M. (1979) A prototype 'on-line' system for management of cotton pests in the Namoi Valley, New South Wales, *Prot. Ecol.*, **1**, 245–64.

Roome, R. E. (1975) Field trials with a nuclear polyhedrosis virus disease and *Bacillus thuringiensis* against *Heliothis armigera* (Hbn.) larvae on sorghum and cotton in Botswana, *Bull. Ent. Res.*, **65**, 507–14.

Roome, R. E. and Daoust, R. A. (1975) Survival of the nuclear polyhedrosis virus of *Heliothis armigera* (Hbn) on crops and in soil in Botswana, *J. Invert. Path.*, **27**, 7–12.

Rose, D. J. W. and Khasimuddin, S. (1979) Wide-area monitoring of the African armyworm *Spodoptera exempta* (walker); (Lepidoptera, Noctuidae), in: R. L. Rabb and G. G. Kennedy (eds), *Movement of Highly Mobile Insects: concepts and methodology in research*, North Carolina State University, North Carolina, pp. 212–17.

Rowell, J. G. (1971) A rapid method of data collection and processing, *J. Agric. Sci.*, **77**, 379–84.

Rudd, W. G., Ruesink, W. G., Newson, L. D., Herzog, D. C., Jensen, R. L. and Marsolan, N. P. (1980) The systems approach to research and decision making for soyabean pest control, in: C. B. Huffaker (ed.), *New Technology of Pest Control*. Wiley Inter-Science, New York, pp. 99–122.

Ruesink, W. G. (1975) Analysis and modelling in pest management, in: R. L. Metcalf and W. Luckmann (eds), *Introduction to Insect Pest Management*. Wiley Inter-Science, New York.

Ruesink, W. G., Shoemaker, C. A., Gutierrez, A. P. and Fick, G. W. (1980) The system approach to research and decision making for alfalfa pest control, in: C. B. Huffaker (ed.) *New Technology of Pest Control*. Wiley Inter-Science, New York, pp. 217–47.

Ruesink, W. G. and Kogan, M. (1975) The quantitative basis of pest management – sampling and measuring, in: R. L. Metcalf and W. H. Luckmann (eds), *Introduction to Insect Pest Management*. Wiley Inter-Science, New York, pp. 309–51.

Russell, G. E. (1978) *Plant Breeding for Pest and Disease Resistance*. Butterworths, London.

Rutherford, I. (1978) The Design and Construction of a Machine to Compare CDA and Hydraulic Nozzles on a Field Scale. *BCPC Monograph* 22, pp. 111–15.

Salt, G. and Hollick, F. S. J. (1944) Studies of wireworm population a: A census of wireworms in pasture, *Ann. Appl. Biol.*, **31**, 53–64.

Sawicki, R. M. (1981) Problems of countering resistance, *Phil. Trans. R. Soc.* London, B. 295, 143–50.

Schopp, R. and Landis, B. J. (1961) A portable sprayer for experimental plots. *J. Econ. Ent.*, **54**, 604–5.

Schuster, M. F. (1980) Insect resistance in cotton, in: M. K. Harris (ed.), *Biology and Breeding for Resistance to Arthropods and Pathogens in Agricultural Crops.* Texas A and M University, pp. 101–12.

Scopes, N. E. A. (1981) Some factors affecting the efficiency of small pesticide droplets, *Proc. 1981 Br. Crop. Prot. Conf. – Pests and Diseases*, 875–82.

Scopes, N. E. A., Randall, R. E. and Biggerstaff, S. M. (1975) Constant temperature, ventilated perspex cage for rearing phytophagous insects, *Lab. Practice*, **24**, 33–5.

Service, M. W. (1970) A battery operated light-trap for sampling mosquito populations, *Bull. Wld. Hlth. Org.*, **43**, 635–41.

Service, M. W. (1976) *Mosquito Ecology: field sampling methods.* Applied Science Publishers, London.

Shepherd, H. H. (1958) *Methods of Testing Chemicals on Insects*, Vol. 1. Burgess, Minneapolis, (306 pp.).

Shepherd, H. H. and Campbell, F. L. (1932) The relative toxicity of rotenone and some related compounds as stomach insecticides, *J. Econ. Ent.*, **25**, 142–4.

Shires, S. W. (1980) Soil surface predators in arable land: the effects of farming practices, *SPAN*, **23**, 62–4.

Shoemaker, C. A. (1980) The role of systems analysis in integrated pest management, in: C. B. Huffaker (ed.), *New Technology of Pest Control.* Wiley Inter-Science, New York, pp. 25–49.

Simanton, W. A. and Trammel, K. K. (1965) Design and performance of a laboratory air-blast sprayer, *J. Econ. Ent.*, **58**, 576–8.

Smith, D. B. and Bouse, L. F. (1981) Machinery and Factors that affect the application of pathogens. In H. D. Burgess, (ed.), *Microbial Control of Pests and Plant Diseases 1970–1980*. Academic Press, New York, 635–653.

Smith, D. B. and Burt, E. C. (1970) Effects of the size of ULV droplets on deposits within cotton foliage both inside and immediately downwind from a treated swath, *J. Econ. Ent.*, **63**, 1400–405.

Smith, D. R., Hostetter, D. L. and Pinnell, R. E. (1980) Laboratory formulation comparisons for a bacterial (*Bacillus thuringiensis*) and a viral (*Baculovirus heliothis*) insecticide, *J. Econ. Ent.*, **73**, 19–21.

Smith, J. G. (1969) Some effects of crop background on populations of aphids and their natural enemies on brussels sprouts, *Ann. Appl. Biol.*, **63**, 326–9.

Smith J. G. (1976) Influence of crop background on aphids and other phytophagous insects on brussels sprouts, *Ann. Appl. Biol.*, **83**, 1–13.

Snedecor, G. W. and Cochran, W. G. (1967) *Statistical Methods*. Iowa State University Press, Iowa.

Soper, R. S. and Ward, M. G. (1981) Production, formulation and application of fungi for insect control, in: G. C. Papavizas (ed.), *Biological Control in Crop Production*. Allanheld, Osmun, London, pp. 161–80.

Sorensen, A. A. and Falcon, L. A. (1980) Microdrop application of *Bacillus thuringiensis*: methods to increase coverage in field crops, *J. Econ. Ent.*, **73**, 252–7.

Southwood, T. R. E. (1966) *Ecological Methods*. Chapman and Hall, London.

Southwood, T. R. E. (1968) The abundance of animals, *Inaug. Lect. Imp. Coll. Sci. Technol.*, **8**, 1–16.

Southwood, T. R. E. (1976) Bionomic strategies and population parameters, in: R. M. May (ed.), *Theoretical Ecology: principles and applications*. Blackwell, Oxford, pp. 30–52.

Southwood, T. R. E. (1979) Pesticide usage, prodigal or precise, *Proc. 1979 Br. Crop Prot. Conf. – Pests and Diseases*, 603–19.

Southwood, T. R. E. (1981) Environmental considerations, *Phil. Trans. R. Soc.* Lond. B. 295, 17–18.

Southwood, T. R. E., Brown, V. K. and Reader, P. M. (1979) The relationship of plant and insect diversities in succession, *Bio. J. Linnean Soc.*, **12**, 327–48.

Southwood, T. R. E. and Norton, G. A. (1973) Economic aspects of pest management strategies and decisions, *Mem. Ecol. Soc. Australia*, **1**, 168–84.

Southwood, T. R. E. and Pleasance, H. J. (1962) A hand-operated suction apparatus for the extraction of Arthropods from grassland and similar habitats, with notes on other models. *Bull. Ent. Res.*, **53**, 125–8.

Spencer, D. M. (1972) Results in practice II: glasshouse crops, in: R. W. Marsh (ed.), *Systemic Fungicides*. Longman, London.

Stapley, J. H. (1973) Insect pests of coconuts in the Pacific region, *Outlook on Agric.*, **7**, 211–17.

Steer, W. (1938) Laboratory methods for the biological testing of insecticides I; Methods of testing ovicides, *J. Pomol.*, **15**, 338–55.

Stent, C. J., Taylor, W. A. and Shaw, G. B. (1981) A method for the production of uniform sized droplets and electrostatic dispersion, *Trop. Pest Management*, **27**, 262–4.

Sterling, W. L. and Peters, E. P. (1979) Sequential decision sampling, in: W. L. Sterling (ed.), *Economic Thresholds and Sampling of Heliothis species*. Southern Co-operative Series Bulletin No. 231.

Stern, V. M., Smith, R. F., van den Bosch, R. and Hagen, K. S. (1959) The integration of chemical and biological control of the spotted alfalfa aphid. I: The integrated control concept, *Hilgardia*, **29**, 81–101.

Stevenson, J. H. (1968) Laboratory studies on the acute and oral toxicities of insecticide to honey bees, *Ann. Appl. Biol.*, **61**, 467–72.

Stevenson, J. H., Needham, P. H. and Waller, J. (1978) Poisoning of honeybees by pesticides: investigations of the changing pattern in Britain over 20 years, in: *Rothamsted Report for 1977, Part 2*, pp. 55–72.

Stokes, R. H. and Robertson, R. A. (1949) Standard solutions for humidity control at 25 °C, *Industr. Engng. Chem.*, **41**, 2013.

Stonebridge, W. C. (1981) Selective post-emergence grass control in broad-leaf arable crops, *Outlook on Agric.*, **10**, 385–92.

Strickland, A. H. (1954) An aphid-counting grid, *Plant Path.*, **3**, 73–5.

Strickland, A. H. (1958) Fritfly attack and yield of oats, *Ann. Appl. Biol.*, **46**, 486–93.

Suter, H. (1978) Effects of pesticides on the beneficial arthropod species *Chrysopa carnea* (steph.) method and results, *Schweiz Landwirtschaft licke Forsehung*, **17**, 37–44.

Symmons, P. and Rosenberg, L. J. (1978) A model simulating mating behaviour of *Spodoptera littoralis, J. Appl. Ecol.*, **15**, 423–37.

Taft, H. M., Hopkins, A. R., Jernigan, C. E. and Webbs, J. C. (1969) A new 8-row ground sprayer with auxiliary air for ULV application of pesticide to cotton, *J. Econ. Ent.*, **62**, 570–4.

Tammes, P. M. L. (1961) Studies of yield losses II: injury as a limiting factor of yield, *Ti'dschr. Pl. Ziekt.*, **67**, 257–63.

Tanton, M. J. (1962) The effect of leaf 'toughness' on the feeding of larvae of the mustard beetle *Phaedon cochleariae, Fab. Entomol. Exp. Appl.*, **5**, 74–8.

Taylor, L. R. (1951) An improved suction trap for insects, *Ann. Appl. Biol.*, **38**, 582–91.

Taylor, L. R. (1955) The standardization of air flow in insect suction traps, *Ann. Appl. Biol.*, **43**, 390–408.

Taylor, L. R. (1960) The distribution of insects at low levels in the air, *J. Anim. Ecol.*, **29**,45–63.

Taylor, L. R. (1962a) The efficiency of cylindrical sticky insect traps and suspended nets, *Ann. Appl. Biol.*, **50**, 681–5.

Taylor, L. R. (1962b) The absolute efficiency of insect suction traps, *Ann. Appl. Biol.*, **50**, 405–21.

Taylor, L. R. (1979) The Rothamsted insect survey: an approach to the theory and practice of synoptic pest forecasting in agriculture, in: R. L. Rabb and G. G. Kennedy (eds), *Movement of Highly Mobile Insects: concepts and methodology in research*. North Carolina State University, North Carolina, pp. 148–85.

Taylor, L. R. and Brown, E. S. (1972) Effects of light trap design and illumination of samples of moths in the Kenya highlands, *Bull. Ent. Res.*, **62**, 91–112.

Taylor, L. R. and Palmer, J. M. P. (1972) Aerial sampling, in: H. F. van Emden (ed.), *Aphid Technology*. Academic Press, London and New York, pp. 189–234.

Taylor, W. A., Merritt, C. R. and Drinkwater, J. A. (1976) An experimental tractor-mounted very low volume uniform drop size sprayer, *Weed Res.*, **16**, 203–208.

Taylor, W. E. and Bardner, R. (1968) Effects of feeding by larvae of *Phaedon cochleariae* (F.) and *Plutella maculipennis* (Curt.) on the yield of radish and turnip plants, *Ann. Appl. Biol.*, **62**, 249–54.

Teich, I., Neumark, S., Jacobson, M., Klug, J., Shani, A. and Waters, R. M. (1979) Mass trapping of males of Egyptian cotton leafworm *Spodoptera littoralis* and large-scale synthesis of procelure, in: F. J. Litter (ed.), *Chemical Ecology: odour communication in animals*. Elsevier Biomedical Press, pp. 343–50.

Teng, P. S. (1981) Use of regression analysis for developing crop loss models, in: L. Chiarappa, (ed.), *Crop Loss Assessment Methods*, Suppl. 3. CAB/FAO, pp. 51–55.

Thompson, A. R., Suett, D. L., Percivall, A. L., Padbury, C. E., Edmonds, G. H. and Farmer, C. J. (1981) *Precision Equipment for Sowing and Treating Small Plots with Granular Insecticide*. Annual Report, National Vegetable Research Station for 1980, pp. 34–5.

Thompson, A. R. and Wheatley, G. A. (1977) Design of trials for evaluating insecticides against some insect pests of vegetables, *Pestic. Sci.*, **8**, 418–27.

Thompson, N. (1982) Meteorology as an aid to crop protection, in: R. B. Austin, (ed.), *Decision Making in the Practice of Crop Protection*. BCPC Monograph 25, 55–63.

Thornhill, E. W. (1974) The adaptation of a stainless steel container for use as a compression sprayer, *PANS*, **20**, 241–5.

Thornhill, E. W. (1978) A motorized insect sampler, *PANS*, **24**, 205–207.

Thornton, M. E. (1974) Modifications of the Oxford Precision Sprayer 1974, *PANS*, **20**, 246–50.

Thornton, M. E. (1977) A portable sprayer for treating experimental plots, *PANS*, **23**, 414–20.

Tinsley, T. W. (1979) The potential of insect pathogenic viruses as pesticide agents, *Ann. Rev. Ent.*, **24**, 63–7.

Tiong, R. H. C. (1982) Oil palm pests in Sarawak and the use of natural enemies to control them, *Proc. Int. Conf. Pl. Prot. in Tropics*, Kuala Lumpur, 363–72.

Tunstall, J. P. (1968) Pupal development and moth emergence of the red bollworm (*Diparopsis castanea* (Hmps.)) in Malawi and Rhodesia, *Bull. Ent. Res.*, **58**, 233–54.

Tunstall, J. P. and Matthews, G. A. (1966) Large-scale spraying trials for the control of cotton insect pests in Central Africa, *Cott. Gr. Rev.*, **43**, 121–39.

Turner, N. J. (1965) A small plot logarithmic dosage sprayer with an internally located concentrate container, *Boyce Thompson Inst. Contrib.*, **23**, 21–2.

Uk, S. (1978) Portable Microtip Nozzle Assembly for Producing Controlled Monosize Droplets for Use in field Microplot Experiments. *BCPC Monograph* 22, pp. 121–7.

Unterstenhofer, G. (1963) The basic principles of crop protection field trials, *Pflanzenschutz-Nachrichten Bayer*, **16**, 81–

van der Graaff, N. A. (1981) Increasing reliability of crop loss information. The use of 'indirect' data, in: L. Chiarappa, (ed.), *Crop Loss Assessment Methods*, Suppl. 3. CAB/FAO, pp. 65–9.

van der Plank, J. E. (1963) *Plant Diseases: epidemics and control.* Academic Press, London and New York.

van Emden, H. F. (1970) Insects, weeds and plant health, *Proc. 10th Br. Weed Cont. Conf. 3*, 953–7.

Van Emden, H. F. (1972) Plant resistance to insect pests – Developing risk-rating methods, SPAN, **15**, 71–74.

van Hamburg, H. and Kfir, R. (1982) Tests of cotton bollworms, *J. Ent. Soc. Sth. Afr.*, **45**, 109–21.

Vickerman, G. P. (1974) Some effects of grass weed control on the arthropod fauna of cereals, *Proc. 12th Br. Weed Cont. Conf.*, **3**, 929–39.

Viggiani, G. and Tranfuglia, A. (1978) Boll. Lab. Ent. Agr. Silvestri, **35**, 8–15.

Voss, G. (1961) A new acaricide test-method for spider mites, *Anz. Schadlinsk*, **34**, 76–77.

Walker, J. K. (1980) Earliness in cotton and escape from the boll weevil, in: M. K. Harris (ed.), *Biology and Breeding for Resistance to Arthropods and Pathogens in Agricultural Crops*. Texas A and M University, pp. 113–23.

Walker, P. T. (1977) Crop losses: some relationships between yield and infestations, *Med. Fac. Landbouww. Rijksui Gent.*, **42**, 919–26.

Walker, P. T. (1981a) The measurement of insect populations, their distribution and damage, in: L. Chiarappa (ed.), *Crop Loss Assessment Methods*, Suppl. 3. CAB/FAO, pp. 43–9.

Walker, P. T. (1981b) Methods for studying the pest/yield relationship and for appraising losses due to insect and mites, in: L. Chiarappa (ed.), *Crop loss assessment methods*, Suppl. 3. CAB/FAO, pp. 73–8.

Walker, P. T. (1983) Crop losses: The need to quantify the effects of pests, diseases and weeds on agricultural production. *Agriculture, Ecosystems and Environment*, **9**, 119–158.

Wall. C., Greenway, A. R. and Burt, P. E. (1976) Electro-anthennographic and field responses in the pea moth *Cydia nigricana*, to sex attractants and related compounds, *Physiol. Ent.*, **1**, 151–7.

Wallace, E. G. R. (1979) A tractor-mounted multiboom system for spray application to small-plot trials, *Proc. 32nd New Zealand Weed and Pest Control Conf.*, 207–10.

Wallace, L. E., McNeal, F. H. and Berg, M. A. (1973) Minimum stem solidness required in wheat for resistance to the wheat stem sawfly, *J. Econ. Ent.*, **66**, 1121–3.

Waller, J. M. (1972) Waterborne spore dispersal in coffee berry disease and its relation to control, *Ann. Appl. Biol.*, **71**, 1–18.

Walton, W. H. and Prewett, W. C. (1949) Atomization by spinning discs, *Proc. Phys. Soc.*, B 62, 341–50.

Wapshere, A. J. (1975) A protocol for programmes for biological control of weeds, *PANS*, **21**, 295–303.

Way, M. J. (1953) The relationship between certain ant species with particular reference to biological control of the Coreid *Theraptus* spp., *Bull. Ent. Res.*, **44**, 669–91.

Way, M. J. (1961) Crop losses by insects and the problem of control, *Proc. Nutr. Soc.*, **20**, 5–11.

Way, M. J. (1973) Objectives, methods and scope of integrated control, in: P. W. Geier, L. A. Clark, D. J. Anderson and H. A. Nix (eds), *Insects: studies in population management.* Ecological Society of Australia Memoirs, Vol. 1, pp. 138–52.

Way, M. J. and Cammell, M. E. (1977) Possibilities and constraints in practical forecasting, *Proc. 1977. Br. Crop Prot. Conf. – Pests and Diseases*, 835–45.

Way, M. J. and Cammell , M. E. (1979) Optimizing cereal yields: the role of pest control, *Proc. 1979 Br. Crop Prot. Conf. – Pests and Diseases*, 633–72.

Way, M. J. and Cammell, M. E. (1981) Effects of weeds and weed control on invertebrate pest ecology, in: J. M. Thresh (ed.), *Pests, Pathogens and Vegetation.* Pitman, London, pp. 443–58.

Way, M. J., Cammell, M. E., Taylor, L. R. and Woiwod, I. P. (1981) The use of egg counts and suction trap samples to forecast the infestation of spring sown field beans *Vicia faba*, by the blackbean aphid *Aphis fabae*, *Ann. Appl. Biol.*, **98**, 21–34.

Way, M. J., Murdie, G. and Galley, D. J. (1969) Experiments of integration of chemical and biological control of aphids on brussels sprouts, *Ann. Appl. Biol.*, **63**, 459–75.

Wheatley, G. A. (1972) *Development of Log-Dose Evaluation Methods.* Report, National Vegetable Research Station for 1971, pp. 66–7.

Wheatley, G. A. and Niendorf, E. K. (1969) *Mechanical Aspects.* Report, National Vegetable Research Station for 1968, p. 75.

Wheatley, P. E. (1962) *Antestia* testing, *Kenya Coffee*, **27**, 405.

Wheatley, P. E. (1963) The giant coffee looper, *Ascotis selenaria reciprocaria, E. Afr. Agric. For. J.*, **29**, 143–6.

WHO (1977) *Programme for Evaluating and Testing New Insecticides, Summary 1975.* WHO/VBC 77.2 (23 pp.).

Wijewardene, R. (1982) Very-low-volume (VLV) knapsack spraying of agrochemicals, *Trop. Pest Man.*, **28**, 170–74.

Wildbolz, J. H. and Meier, W. (1973) Integrated control: critical assessment of cases histories in affluent economics, in: P. W. Geier, L. A. Clark, D. J. Anderson and H. A. Nix (eds), *Insects: Studies in Population Management.* Ecological Society of Australia, Canberra, pp. 221–31.

Williams, C. R. (1948) The Rothamsted light trap, *Proc. R. Ent. Soc.*, London, A. 23, 80–5.

Williams, C. R. (1952) *How Far Do Insects Travel*? Rothamsted Experimental Station, Report for 1951, pp. 172–180.

Wilson, A. G. L. (1982) Past and future *Heliothis* management in Australia, in: W. Reed, (ed.), *Proceedings of the International Workshop on* Heliothis *management.* ICRISAT, Patancheru, A.P. India, pp. 343–54.

Wilson, B. J. (1981) Techniques for the assessment of *Avena fatua* (L.) *Grass weeds in cereals in the UK Conf. 1981*, 93–100.

Wilson, B. J. and Cussans, G. W. (1978) The effects of herbicides applied alone and in sequence, on the control of wild oats *Avena fatua* and broadleaved weeds, and on yields of winter wheat, *Ann. Appl. Biol.*, **89**, 459–66.

Wodagenah, A. and Matthews, G. A. (1981) The addition of oil to pesticide sprays: effects on droplet size, *Trop. Pest Man.*, **27**, 121–4.

Woke, P. A. (1938) The biological disposition of rotenone after ingestion by the southern armyworm, *J. Agric. Res.*, **57**, 707–12.

Wolf, R. W. (1961) Study of the vibrating reed in the production of small droplets and solid particles of uniform size, *Rev. of Sci. Instrum.*, **32**, 1124–9.

Wolfe, M. S. (1981) Integrated use of fungicides and host resistance for stable disease control, *Phil. Trans. R. Soc.*, London B. 295, 175–84.

Wooten, N. W. and Sawyer, K. F. (1954) Pick up of spray droplets by flying locusts, *Bull. Ent. Res.*, **45**, 177–97.

Worthing, C. (ed.) (1983) *The Pesticide Manual.* 7th Edition BCPC.

Wright, M. E. and Standifer, L. C. (1969) Development of a logarithmic attachment for field sprayers, *Proc. Southern Weed Conf.*, 1969.

Yates, W. E. and Akesson, N. B. (1978) Monodisperse atomization systems for pesticide sprays, *Ist Int. Conf. on Liquid Atomization Systems*, Fuel Society of Japan, Tokyo 181–5.

Yates, W. E. and Ashton, F. M. (1960) Logarithmic dosage sprayer, *Agric. Eng.*, **41**, 436–8.

Zadoks, J. C. (1981) EPIPRE, a disease and pest management system for winter wheat developed in the Netherlands, *EPPO Bull.*, **11**, 365–9.

Zadoks, J. C., Chang, T. T. and Konzak, C. F. (1974) A decimal code for the growth stages of cereals, *Weed Res.*, **14**, 415–21.

Zandstra, B. H. and Motooka, P. S. (1978) Beneficial effects of weeds in pest management: a review, *PANS*, **24**, 333–8.

Index